2/6/93

FUNDAMENTALS
OF
CAD

SECOND EDITION

FUNDAMENTALS
OF
CAD

SECOND EDITION

GARY R. BERTOLINE

**ASSISTANT PROFESSOR OF ENGINEERING GRAPHICS
THE OHIO STATE UNIVERSITY
COLUMBUS, OHIO**

 DELMAR PUBLISHERS INC.®

NOTICE TO THE READER

Publisher does not warrant or guarantee any of the products described herein or perform any independent analysis in connection with any of the product information contained herein. Publisher does not assume, and expressly disclaims, any obligation to obtain and include information other than that provided to it by the manufacturer.

The reader is expressly warned to consider and adopt all safety precautions that might be indicated by the activities described herein and to avoid all potential hazards. By following the instructions contained herein, the reader willingly assumes all risks in connection with such instructions.

The publisher makes no representations or warranties of any kind, including but not limited to, the warranties of fitness for particular purpose or merchantability, nor are any such representations implied with respect to the material set forth herein, and the publisher takes no responsibility with respect to such material. The publisher shall not be liable for any special, consequential or exemplary damages resulting, in whole or in part, from the readers' use of, or reliance upon, this material.

This book is dedicated to my parents,
Robert and Caroline, for their constant love,
guidance, and understanding.

Cover photos courtesy of Control Data

Chapter opening art:
Chapters 1, 2, 4, 6 appear courtesy of VersaCAD Designer Corp.
Chapter 3 appears courtesy of Hewlett-Packard
Chapters 4, 8, 10 appear courtesy of CADKEY
Chapter 5 appears courtesy of Autodesk, Inc.
Chapters 7 and 11 appear courtesy of Matra Datavision
Chapter 9 appears courtesy of Robo Systems Corp.

For information address Delmar Publishers Inc.,
2 Computer Drive, West, Box 15-015
Albany, New York 12212-5015

Associate Editor: Joan Gill
Production Editor: Christopher Chien
Production Coordinator: Linda Helfrich
Design Coordinator: Linda Johnson/Susan Mathews

Printed in the United States of America
Published simultaneously in Canada
by Nelson Canada,
A division of The Thomson Corporation

10 9 8 7 6 5 4 3

Library of Congress Cataloging in Publication Data

Bertoline, Gary R.
    Fundamentals of CAD / Gary R. Bertoline.
        p.   cm.
    Bibliography: p.
    Includes index.
    ISBN 0-8273-3291-2.   ISBN 0-8273-3292-0 (instructor's guide)
    1. Engineering design—Data processing.   2. Computer-aided design.
I. Title.
TA174.B454 1988
620'00425'02854—dc19   87-25948
                                                          CIP

CONTENTS

SECTION 3   CREATING CAD DRAWINGS

SECTION 4   3D CAD, PRESENT AND FUTURE APPLICATIONS

SECTION 5   APPENDIX

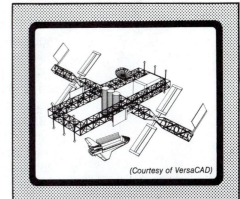

(Courtesy of VersaCAD)

PREFACE

Human beings have always sought the use of tools to make life easier and to enable them to become more productive. Certain tasks associated with drafting-design have always been tedious and time consuming. Thus, people have recently turned to the computer as a means of easing these tasks and becoming more productive in many different applications, including drafting-design.

Thomas Carlyle, the nineteenth century Scottish historian, wrote, "Man is a tool-using animal; without tools he is nothing; with tools he is all." CAD is simply the latest *tool* used in drafting-design to make life easier and the work more productive. As with any tool, the operator must understand that tool and be able to utilize it to its full potential in order for it to be useful. CAD is a very powerful tool for the designer, but it is no more than an expensive toy if it is not used to its full potential. This is a comprehensive text that explains CAD's development, components, future direction, operation, and applications, and how CAD fits into the factory of the future.

The *Fundamentals of CAD*, by Gary R. Bertoline, is intended for students enrolled in industrial technology schools, engineering technology schools, and vocational-technical schools. The book is designed to supplement current drafting texts which are used to teach the fundamentals of drawing. No one should attempt to use a CAD system without a basic knowledge of drafting. The necessary prerequisite skills include orthographic projection, axonometric projection, sectional drawings, dimensioning, geometric construction, and basic math. The text may also be used by people who are interested in learning about CAD, but not necessarily in learning how to use it.

The approach used in this text to teach CAD is the same as it is for teaching drawing using traditional tools. Very simple tasks, such as drawing random lines of any length or angle, are used to reinforce the first CAD drawing commands taught. Each new command taught is reinforced with simple drawing tasks, followed by exercises that are more complicated. Additional drawing assignments from a traditional drafting text can be used to supplement the drawings used in this book. Where applicable, chapters have prerequisites, objectives, a glossary, a list of drawing commands along with their definitions, a review, and drawing exercises.

To make the text relevant to the wide range of CAD systems that are on the market, it has been written to be used in conjunction with the manual of the CAD system on which the student is training, by explaining each drawing command in generic terms. In this way, trainees learn on their own, and will retain this knowledge longer than they would by reading the exact list of steps necessary to perform a task. An amazing amount of similarity exists among systems and the way in which certain drawing tasks are implemented. The procedures to follow may not be exactly the same for each system, but there is enough similarity among most functions that a generic explanation will be useful to the reader. This general knowledge will help trainees to better understand their own systems.

The text is divided into five major sections. The first section, consisting of Chapters One through Three, introduces the reader to computers as a means to better understand CAD. The second section, consisting of Chapters Four and Five, explains in detail the components and basic operation of a typical CAD system. Section 3, consisting of Chapters Six through Nine, explains how CAD is used to create drawings. This is accomplished with illustrations and detailed steps. Chapters Ten and Eleven in Section 4 explain 3-D drawings and applications of CAD, as well as how CAD fits into the automated factory. The fifth and final section consists of the Appendix. The Appendix includes a cross section of CAD vendors with brief descriptions, some points to be considered when purchasing a system, and various sources for learning more about CAD.

The use of color in CAD can dramatically add to a viewer's visualization of a design. At one time, displaying graphics in color was a very expensive alternative to monochrome displays. Now, with the lower costs associated with color graphics, color is being used by a growing number of CAD systems. A section of color photos is included in the text to enable the reader to visualize the tremendous power of color in displaying graphics.

This insert shows the reader how color can be used to enhance such various applications of CAD as mechanical, architectural, electronic, business graphics, and plant layout. Wireframe images of parts, and the finite element displaying stress levels in color is also shown. Three-dimensional models of parts showing shaded surfaces are included in a sequence of photos demonstrating the design of a product on a CAD/CAM system, from the initial design of the model through the machining of the part. Additional photos are included to show other applications of CAD to give the reader an appreciation for the tremendous potential of CAD/CAM.

This second edition is a result of user feedback and a general updating of information. CAD is evolving rapidly and changes in hardware, software, and capabilities seem to occur daily. When the first edition was being written, the use of microcomputers for CAD had just begun. This edition has been updated and includes more information on microcomputer-based CAD.

Chapters One through Five include updated information on hardware and software, as well as additional information about new technologies being used with CAD. Chapters Six through Ten contain additional commands not included in the first edition. The section on plotting has been moved from Chapter Nine to Chapter Six, and material originally placed near the end of the text has been moved to Chapter Seven.

Other major changes have been made in this edition: New and more powerful color photographs have been included; an eleventh chapter, which contains some information from Chapter Ten in the first edition, has been added to the text. (This was necessitated by the addition of 3-D drawings in Chapter Ten.) Finally, the number of drawings to be used for assignments has been increased. Many of these drawings are more challenging and can be used as design projects by the students. I hope that you find these changes to be improvements that strengthen this text and provide you or your student with the fundamentals of CAD.

ACKNOWLEDGMENTS

I am grateful to the many people who helped to make this book possible. To Don Wellman and John Hawley for their review of the manuscript and suggested additions and revisions. To Dan Ranly and Kevin Gilliland at STAMCO, Tony Hemmelgarn at CAD/CAM, Incorporated, and Debra Sattler from Bruning CAD for their technical expertise. To Bob Harrell, Randy Evers, Tom Tremper, Craig Miller, Angela Reath, and David Brackman for their drawings included in the text. To all the students who suffered through the early drafts of the manuscript and assisted in "debugging" it. To Muriel Pellegrino for her typing, copying and patience. To Mark Huth at Delmar, and the following reviewers who assisted the author and the publisher with the development of the manuscript.

J. David Alpert
Santa Ana College

James Hysaw
Oklahoma City Community College

Van Nichols
Chemeketa Community College

Pat Peper
Auerbach Publishers, Inc.

Richard J. Svoboda
Muskegon Community College

James R. Vandervest
Gulf Coast Community College

James R. Woughter
State University of New York, Alfred

Finally, to my family for their faith and understanding during the many hours that this project took my time from them. This is especially true of my wife, Ada, whose proofreading, typing, patience, understanding, and midnight snacks fulfilled my need for sustenance and inspiration.

I am sincerely interested in your comments and suggestions for improving this textbook. Address them to:

Gary Bertoline
Department of Engineering Graphics
240 Hitchcock Hall
Ohio State University
Columbus, OH 43210

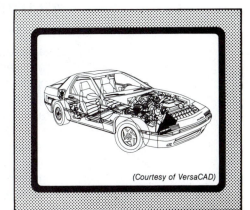

(Courtesy of VersaCAD)

SECTION 1
INTRODUCTION TO CAD

Chapter One
CAD and the Drafting-Design Field

Computer-aided drafting (CAD) has revolutionized the drafting-design field. CAD has found its way into industry, changing the methods used to produce drawings. The basic tools of the trade are being replaced by the computer. Compasses, triangles, and pencils are being replaced by cathode ray tubes (CRTs), mice, and plotters.

This does not mean that learning the basic skills of drawing and using the traditional tools of the trade will be eliminated, or that a drafter trained on the board will be replaced by a computer operator. The computer is simply another tool used by the drafter to produce drawings. Although CAD is a very powerful tool, drafters trained in the fundamentals of drafting are necessary to operate a CAD system efficiently.

Today, a well-trained drafter preparing for the job market should be exposed to the basics of CAD. This chapter begins by reviewing the traditional methods of producing drawings. It goes on to explain why computers are being used for drafting, and then defines CAD and looks briefly at CAD today and the factory of the future.

OBJECTIVES

After completing this chapter, you will be able to

- explain the reasons for the use of computers in drafting.
- define *computer-aided drafting* and *computer-aided design*.
- list some common uses of CAD.
- explain the effects that CAD has on the drafting-design occupation.
- discuss industry expectations for a CAD operator.
- list the advantages of using CAD over traditional drafting techniques.
- define the terms *hardware* and *software*.

TRADITIONAL TOOLS OF THE TRADE

The traditional tools of the trade have not changed drastically since their introduction. The major changes in the last century have been in methods of reproduction, drawing media, and the use of drafting machines to replace the T-square or parallel edge. Until CAD, drafters were using the same tools to draw as were used by Euclid and Pythagoras: a compass, a straightedge, and a pencil. These traditional tools have been replaced with the development of CAD. Figure 1-1 shows an office having CAD workstations in place of drafting boards and traditional tools.

The total elimination of traditional drafting tools will not occur for some time to come, but the day when the majority of industries have CAD systems is at hand. The need to learn the traditional methods of drawing and good graphic skills is still necessary, but may be emphasized less. As CAD technology improves and prices continue to drop, making these tools more common, the need for CAD training will grow. Further growth of CAD can be expected due to the numerous advantages of CAD over traditional drafting methods, industry's need to become more productive, and the need to eliminate repetitious work for the designer/drafter.

FIGURE 1-1 Drafting office using CAD terminals in place of traditional drafting tools. *(Courtesy of Applicon)*

THE ADVANTAGES OF CAD

Occasionally, drafting can be a repetitious, inefficient, and time-consuming task. Lettering drawings by hand and making changes to them are two examples of this. CAD can make a drafter much more efficient. Depending on the drafting task, two to ten times production improvement is not uncommon when using a CAD system. The following are the advantages of using CAD over manual drafting.

Speed

In almost all areas of drafting, CAD is faster than manual drawing. This is especially true as the operator becomes more proficient through continued use of the system.

Accuracy

Dimensions are keyed-in using the computer keyboard or menu instead of reading a scale. Mating parts can be checked for fit by having the computer match the parts on the screen before hard copies are produced.

Revisions

One of the most time-consuming tasks that a drafter performs is making changes to existing drawings. With a CAD system, revisions are much faster than manual methods. After a drawing has been completed, it is stored and can be recalled at any time to make changes easily. The revised drawing can be stored so that both versions of the drawing will now be in memory for later use. Also, alternative designs are much easier to produce, making the designer more creative and reducing the time between design and production.

Neatness

After the drawing has been produced on screen, it can be drawn on paper with a plotter. The mechanical plotting of a drawing produces clean, accurate, and neat drawings using proper line weight and consistency, and sharp, consistent lettering.

Legibility

Mechanical plotting can also produce a drawing that is easier to read and understand. This is accomplished by using different pen thickness and colors, along with different lettering styles. Figures 1-2 and 1-3 show the difference between a manually produced drawing and one produced with a CAD system.

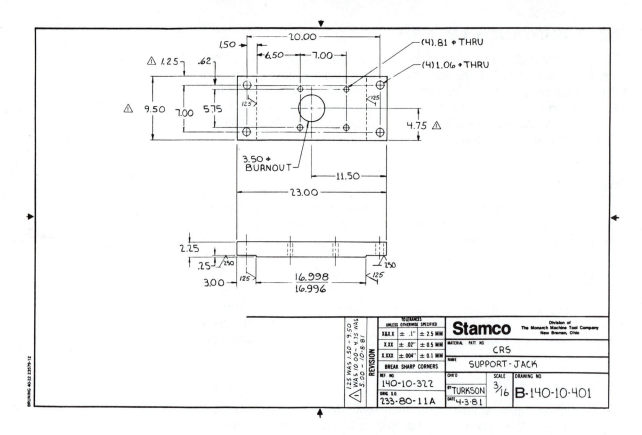

FIGURE 1-2 Drawing produced using traditional tools. *(Courtesy of STAMCO Division of the Monarch Machine Tool Co.)*

Cost

In the past, CAD systems were an expensive alternative to manual drafting, and only major industrial corporations could afford them. Now, because of reduced costs of memory, increased competition, and improved software and computer technology CAD systems are affordable to most drafting-design offices.

Repetition

A drafter spends a great deal of time on tedious work, such as lettering, line weight and consistency, and tracings. The use of CAD can make the drafter more productive and improve the drafting occupation by reducing this tedious and repetitious work.

With CAD, lettering is as easy as keying-in the letters from a keyboard. Line weight and consistency are controlled by the operator and flawlessly plotted by the computer. Templates are used in drafting for common repetitive features. Templates, in the form of symbols, can be created with CAD systems for both common and uncommon features. Simple symbols, such as those shown in Figure 1-4, or more complex ones as shown in

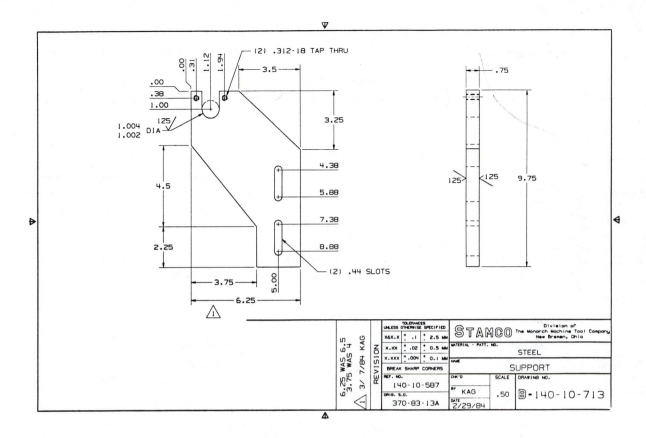

FIGURE 1-3 Drawing created and plotted using CAD. (*Courtesy of STAMCO Division of the Monarch Machine Tool Co.*)

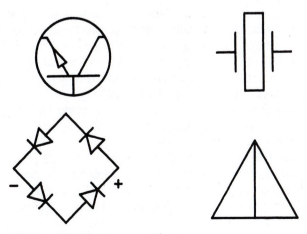

FIGURE 1-4 Electronic symbols drawn with CAD. Once created, these symbols can be located, scaled, and rotated into position on the drawing.

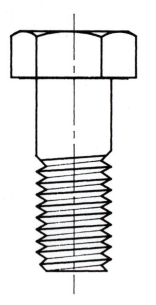

FIGURE 1-5 A more complicated symbol produced with CAD.

Figure 1-5, are created by drawing them with the CAD system and storing them on disk to be used at any time. Once a drawing has been produced and stored, it does not have to be drawn again if it is part of another drawing. Parts of previous drawings can be combined to produce new drawings. Memory allows such information as drawings or programs to be stored in the computer.

Variety of Functions

A *function* is a group of commands that enables the CAD system to perform a specific task. For example, a mechanical drafting CAD system has a number of drawing functions such as lines, arcs, and dimensions that can be used to produce a mechanical drawing. Most CAD systems are capable of producing virtually anything that can be produced on a drafting board. CAD can be used in practically every drafting discipline, such as mechanical, architectural, electronic, and piping, as well as others.

WHAT IS CAD?

Throughout this textbook, the abbreviation CAD appears many times, and is used to refer to computer-aided drafting. *Computer-aided drafting* is the use of computers, software and associated hardware to produce drawings that would normally be prepared manually. It is automated drafting, commonly executed on a microcomputer. *Software*, commonly stored on disk, is the chained statements, directions or procedures used by the computer to perform a task. *Hardware* is all the physical equipment or devices associated with the operation of a computer.

CAD is sometimes referred to as computer-aided design. This meaning can be different from the meaning for computer-aided drafting. The major difference between the two systems is in the software and the hardware. A computer-aided design system can perform automated drafting but, in addition, it also uses the computer for designing and analyzing. A design system is capable of designing and analyzing through the use of computer graphics, usually peformed on mini- or mainframe computers. A design system typically has design analysis functions such as finite element analysis, mass properties calculations, and 3-dimensional (3-D) drawing capabilities. This does not mean that a drafting system could not have these capabilities. It means only that these engineering functions are more common on a computer-aided design system. To repeat, CAD in this textbook usually refers to computer-aided drafting.

A typical CAD system consists of a central computer, a workstation that displays the drawing on a monitor, a keyboard to input typed information, a cursor control device, and a plotter and/or hard-copy device to produce drawings on paper. Figure 1-6 shows a typical CAD system.

FIGURE 1-6 Typical CAD workstation showing the computer below the desktop, workstation, input device, and hard-copy device and plotter. *(Courtesy of Gerber Systems Technology, Inc.)*

How a Typical CAD System Works

Using an input device to locate points on the monitor, a drawing can be quickly created. The computer draws straight lines, perfect circles and arcs, differing line types and thicknesses, various crosshatching patterns, and irregular curves. The computer can be called upon to position and draw standard components in much the same way a template is used to save time drawing identical features when using traditional tools. The drawing scale can be quickly changed by keying-in the new scale. The computer will make all the necessary calculations and redraw the part to the new scale. One can easily zoom in on an object to do detail work and quickly zoom out to view the drawing in full. Crosshatching for sectional views is done by identifying the area to be crosshatched and the area will be instantly filled with crosshatching. The crosshatching symbol, spacing, and angle can be set by the drafter or chosen from preprogrammed styles. Line types can also be easily controlled and changed by the operator.

Once a drawing is complete, dimensions are automatically calculated by the computer and placed anywhere on the drawing by the operator. Notes and labels of numerous styles and sizes can be typed in and placed on the drawing. A multicolored plot can then be made of the completed drawing. A typical CAD system has the potential to produce virtually any drawing function that can be done on a board.

COMMON APPLICATIONS OF CAD

A CAD workstation can have many different drawing applications. Software is available that will produce many types of drawings. Some of the common drawing applications of CAD include: mechanical, architectural, electrical, electronic, piping, and civil. Figures 1-7 through 1-17 show some common application drawings produced with CAD. CAD is not limited to those applications shown in the figures. Software is available for producing virtually any type of drawing needed by industry.

THE EXPECTED GROWTH OF CAD

Ever since computers have been used to produce drawings, there has been a steady growth in the use of CAD by industry. The early 1960s saw the beginning of CAD. The state of the technology then made it available only to the largest corporations, such as General Motors and Boeing. The early 1970s saw the beginning of the first turnkey CAD systems available on the market along with a drop in memory costs due to advanced computer technology. Thus, CAD became more affordable to many companies. The early 1980s saw a dramatic increase in the number of vendors marketing standalone turnkey CAD systems. Continuing technological advances and increased competition in the market caused a further drop in price.

The worldwide computer graphics market in 1983 amounted to $3 billion. Industry growth in the early 1980s was about 30%, and is expected to continue to grow at a 15% rate through 1990 to an estimated market of $8–9 billion. For the drafter, this means that the drafting occupation will rapidly change from board work to CAD operator. This does not mean that a CAD operator will not need prior training in the traditional manner. The concepts learned in the traditional drafting-design curriculum are still essential in becoming a successful CAD operator. To prepare a drafter for CAD, all that is needed in addition to a traditional drafting-design degree is a course teaching the fundamentals of CAD and/or other design courses integrating CAD.

As CAD increases in popularity, it will become the core of the curriculum. Skills learned on one particular CAD system can be quickly transferred to another system. Thus, the type of system on which one is trained is not as important as the necessity of having CAD training to prepare for the expected growth of CAD in industry.

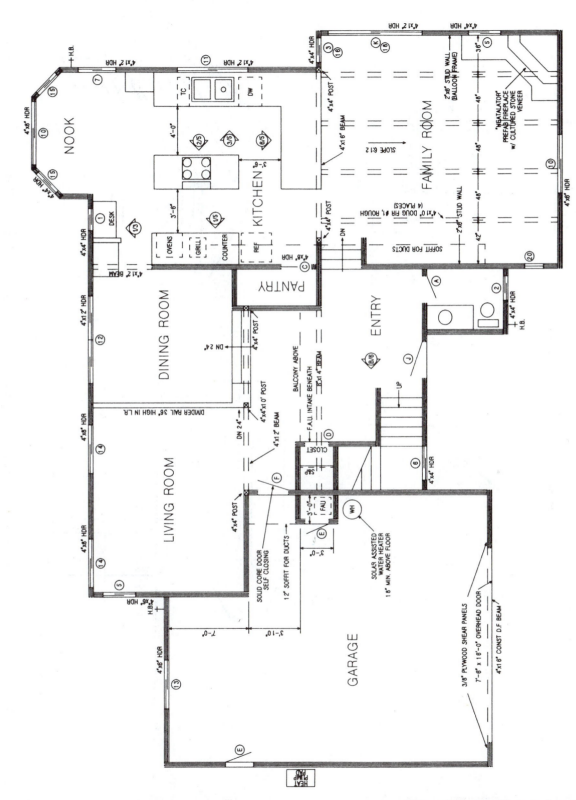

FIGURE 1-7 Architectural drawing produced with CAD. (Courtesy of Hewlett-Packard)

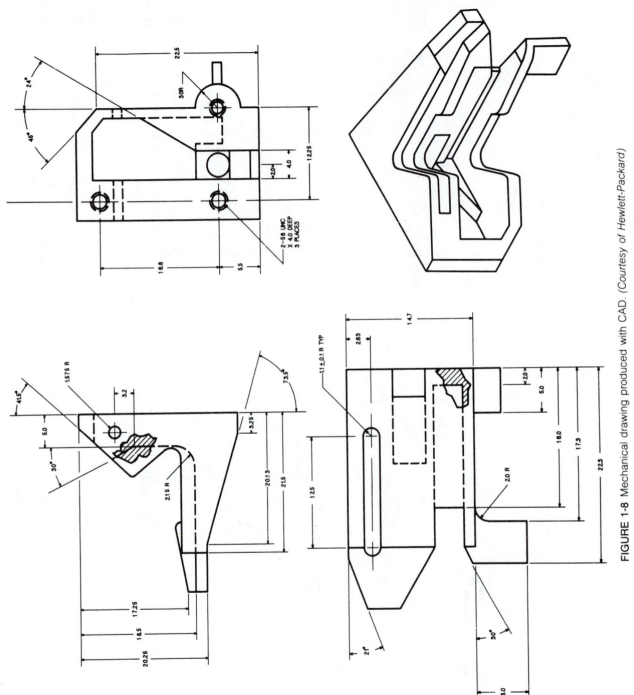

FIGURE 1-8 Mechanical drawing produced with CAD. (Courtesy of Hewlett-Packard)

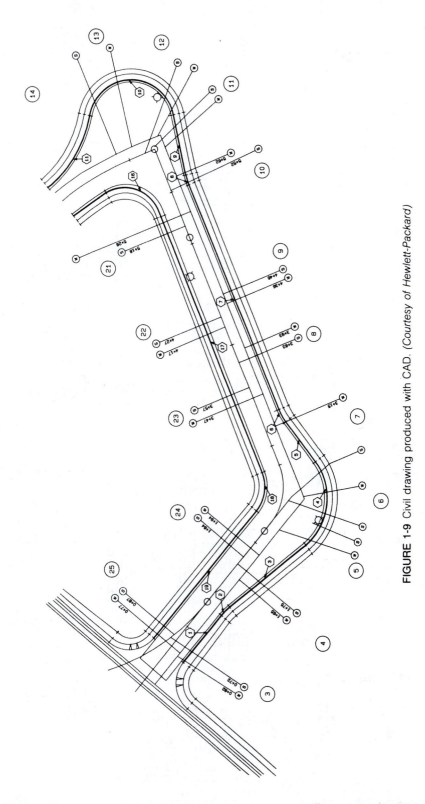

FIGURE 1-9 Civil drawing produced with CAD. (Courtesy of Hewlett-Packard)

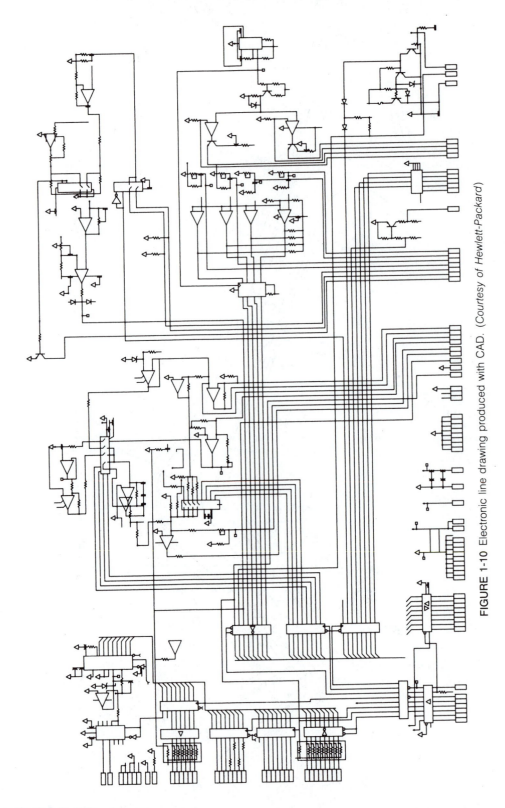

FIGURE 1-10 Electronic line drawing produced with CAD. (Courtesy of Hewlett-Packard)

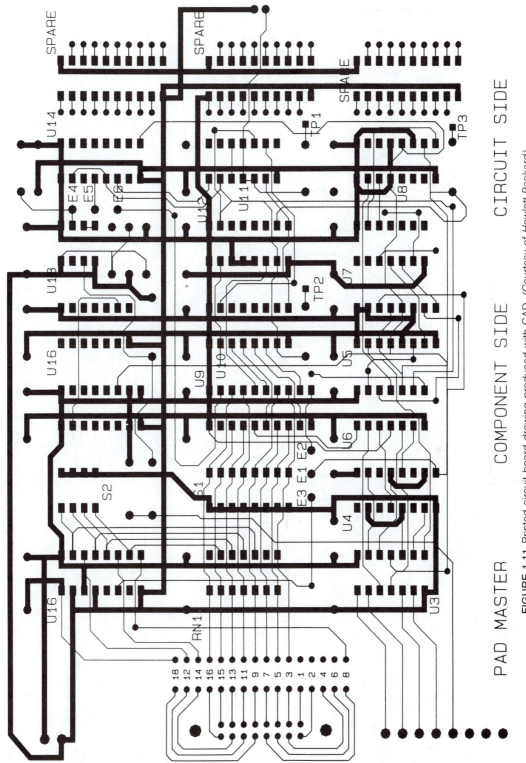

PAD MASTER   COMPONENT SIDE   CIRCUIT SIDE

FIGURE 1-11 Printed circuit board drawing produced with CAD. (Courtesy of *Hewlett-Packard*)

FIGURE 1-12 Mapping drawing produced with CAD. (*Courtesy of Hewlett-Packard*)

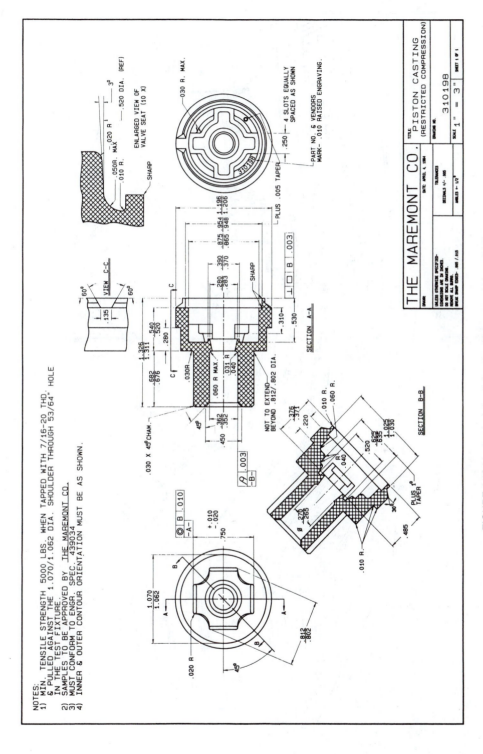

FIGURE 1-13 Mechanical detail drawing (*Courtesy of VersaCAD*)

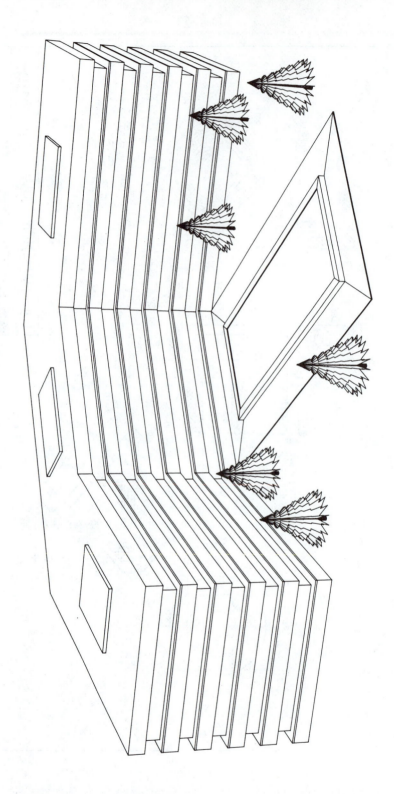

FIGURE 1-14 3-D drawing produced on a Microcomputer-based CAD system (*Courtesy of VersaCAD*)

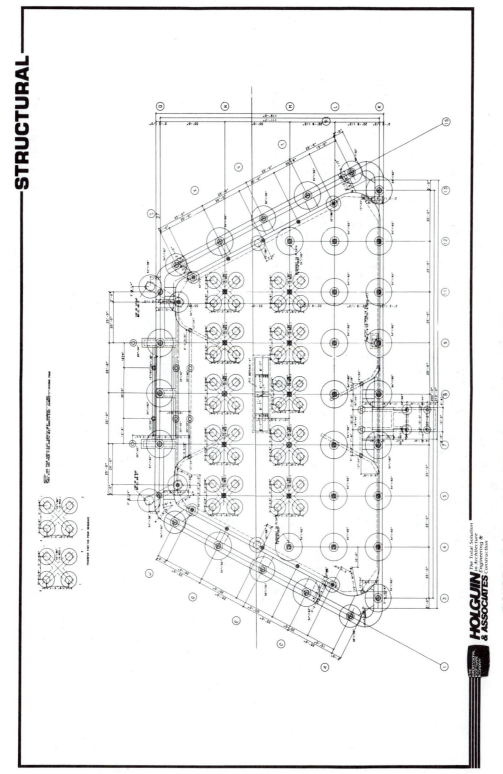

FIGURE 1-15 Structural drawing produced with CAD. (Courtesy of Holguin & Associates, Inc.)

CAD and the Drafting-design Field • 17

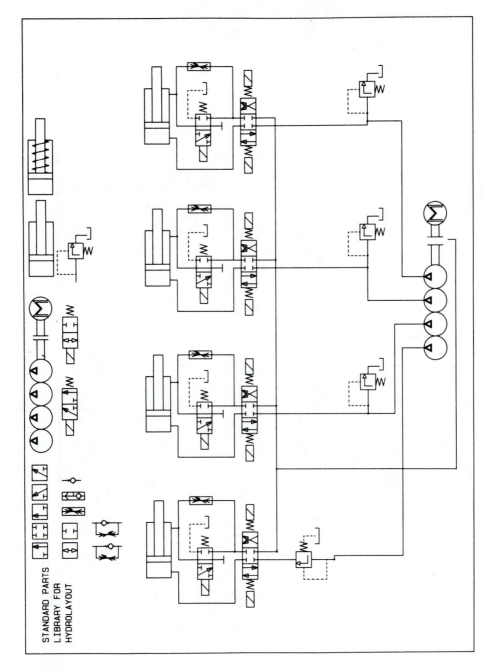

FIGURE 1-16 Hydraulic system produced with CAD using a standard parts library shown at the top of the drawing. *(Courtesy of Hewlett-Packard)*

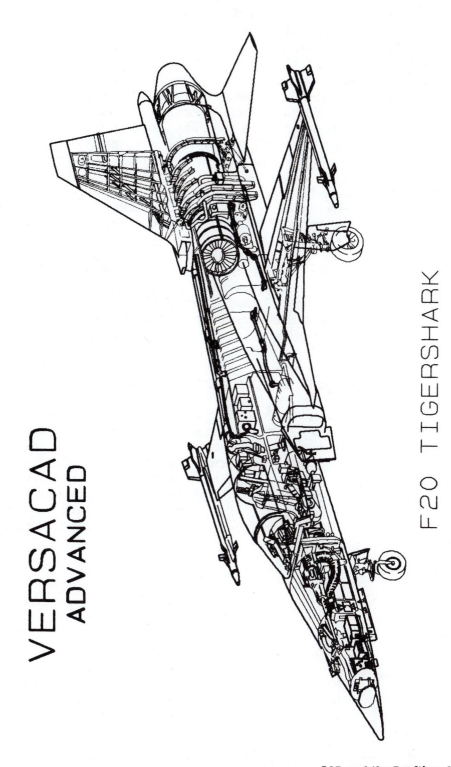

VERSACAD ADVANCED

F20 TIGERSHARK

FIGURE 1-17 CAD can also be used to create technical illustrations. *(Courtesy of VersaCAD)*

The Effects of Computers on Design/Drafting

The advent of computers in design/drafting will have a definite effect on related occupations and the workplace. Drafters have always had to contend with such tedious tasks as lettering and striving for line weight consistency. CAD can help to alleviate some of the tedium associated with drafting. The drafter can now become more productive and creative. Advanced software gives the drafter an opportunity to perform more challenging tasks.

Several negative effects are associated with CAD. Some individuals are apprehensive about using computers. This fear can be a real hindrance to learning and adjusting to computers in drafting. Unless this fear is conquered, the drafter will not become a very efficient CAD operator. A good CAD operator has no fear of the computer, and should have an attitude that will foster improvement in technique. A CAD system can make a drafter much more productive, thus leading to a more challenging and fulfilling career. If CAD training is approached in this manner, learning to operate a CAD system will be a positive experience.

Two other negative effects of CAD deal with shift-work and downtime. The CAD operator has little control over either of these problems. Because of the high initial cost of CAD, some industries have gone to shift-work to keep the CAD system running 24 hours a day to reduce payback time. *Downtime* is associated with a computer breakdown due to hardware or software problems. Downtime can cripple a drafting department if the problem is serious and service is not immediate.

Eventually, the emphasis on drawing and rendering skills may change to skills of problem solving, fundamental design, and judgment. Drafters will not be replaced by computer operators unskilled in drafting concepts. These concepts are important because, for example, the computer cannot place dimensions, section a view, or determine proper drafting practices or standards without human intervention. This human intervention should come from a person trained in drafting-design who knows these concepts. That trained drafter, with CAD experience, will then be able to operate the system at maximum efficiency. This will result in increased productivity and a shorter payback time for the capital investment, which is a major concern of any company contemplating the purchase of a CAD system.

Industry Expectations

Industry is constantly looking for ways to increase productivity. For this reason alone, industry can be expected to adopt CAD quickly. This is also the reason that skilled drafters will be needed to operate CAD systems. Industry seeks drafters with the initiative to expand on the system and do other more advanced problem-solving work. Programming skills are not required, but they are helpful with systems capable of macros or parametric programming routines. *Parametric programs* are English or languagelike statements that can be chained together to perform a task. Having some programming background would assist the CAD operator in writing these powerful and time-saving programs. Finally, typing skills may become more important to drafters because of the large amount of typing that is associated with CAD. Although it is not required, CAD operators may find themselves to be more efficient if they have taken a course in typing.

CAD TODAY AND THE FACTORY OF THE FUTURE

Computers will have a tremendous effect on our basic manufacturing industries. The computer is already extensively used in drafting-design. However, the computer's role in manufacturing does not stop with the design and drawing of the product. The actual manufacturing of the part can be controlled by the computer. The process of computer control of such manufacturing operations as welding, drilling, milling, and so forth, is called *computer-aided manufacturing* (CAM). CAM can also mean the controlling of robots to assist in manufacturing operations and such automatic inspection devices as sensor robots and coordinate measuring machines (CMM).

CAM is used when a number of parts have to be manufactured exactly alike. Because each part is identical, a program can be made to control machine movements. This has been done for a number of years with numerical control (NC) machine tools. A more recent development has been the introduction and use of computers to assist in programming NC machine tools.

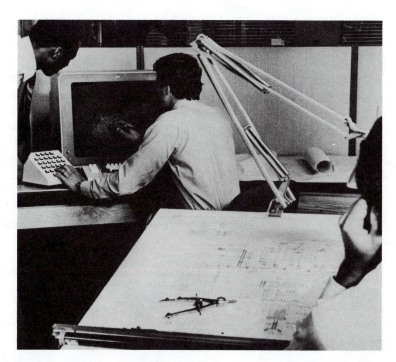

FIGURE 1-18 In the future, CAD will continue to replace or supplement traditional drafting tools. *(Courtesy of International Business Machines Corporation)*

CAD and CAM have many distinct advantages in the manufacturing of a product. Joining these two operations together is the ultimate goal in manufacturing. This combined operation is called CAD/CAM, and it involves a designer for creating the part on screen, and a designer or NC parts programmer for designating the tool path. A program is then generated and punched onto tape to control the machine tools used to manufacture the part. The CMM can then be programmed to automatically inspect the manufactured parts.

As computer technology advances and CAD/CAM software is improved, the control of manufacturing will be accomplished more and more by computers. CAD will be used increasingly in the years to come as manufacturing moves towards automation. As automation increases, the need for trained CAD operators will also increase. It is estimated that more than 1 million new jobs for CAD/CAM workers will be created by the year 2000.

The ultimate effect of automated manufacturing and CAD on the drafting-design occupation is difficult to predict. As CAD becomes the dominant method of producing drawings, the emphasis on drafting technique will diminish. Drafters trained in the traditional manner will need additional training in CAD. As computers become an integral part of our society, dramatic changes will occur both on the job and in the home. The drafter is not immune to these expected changes brought on by the computer. A well-trained drafter must have CAD training to be able to compete on the job market today and in the future.

Chapter One GLOSSARY

Computer-aided design—the use of a computer, software, and associated hardware to produce drawings as well as to perform complex engineering functions.

Computer-aided drafting—the use of a computer, software, and associated hardware to produce drawings.

Downtime—the period of time associated with a computer breakdown caused by hardware or software problems.

Function—a group of commands that enables a CAD system to perform a specific task, such as drawing circles.

Hardware—all the physical equipment or devices associated with the operation of a computer.

Software—the chained statements, directions or procedures used by the computer to perform a task. Sometimes the disks used to store programs are referred to as software.

Chapter One REVIEW

1. Why are basic drafting techniques still of value to a CAD operator?
2. Why are computers being used in drafting?

3. Explain some of the positive and negative effects of CAD on the drafting occupation.
4. What are some of industry's expectations for a CAD operator?
5. List five common drawing applications for which CAD can be used.
6. List five advantages of CAD over traditional drafting methods.
7. Explain briefly the CAD/CAM process.
8. Explain the difference between computer-aided drafting and computer-aided design.
9. Explain the difference between hardware and software.

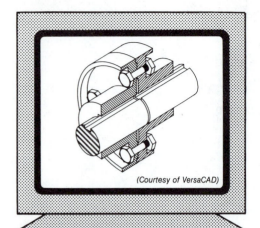

(Courtesy of VersaCAD)

__ **Chapter Two** __

The Development of Computer Graphics

The history of computer graphics relates closely to the development of the computer. Early methods of generating graphics with computers were slow and not very refined, just as the first computers were slow, bulky, hard to program, and expensive. But as computer and electronic technology advanced and hardware devices became more sophisticated, computer graphics became more refined. Research undertaken at universities and major corporations led the way in the development and advancement in computer graphics.

This chapter gives the reader a brief overview of the development of the computer and the corresponding developments in computer graphics. This historical review will give you a feel for the rapid growth in computer graphics. The necessity for traditional drafter/designers to receive CAD training will become more apparent as we look at the rapid growth of CAD expected in the future. Because of the rapidly changing technology, competition, and falling prices, CAD systems are becoming the norm rather than the exception in a design/drafting office.

OBJECTIVES

After completing this chapter, you will be able to

- list reasons for the sudden growth in computer graphics.
- explain some of the future developments expected in computer graphics.
- discuss the development of computer graphics.
- describe the difference between an analog computer and a digital computer.
- explain how computers have been grouped by generations.
- define *interactive computer graphics*.
- define *microprocessor* and describe its function.

THE DEVELOPMENT OF THE COMPUTER

The development of the electronic computer began in the late 1930s and early 1940s. Computers developed along two lines: the analog computer and the digital computer. An *analog* computer uses variations in physical quantities such as electrical voltage to process data. This type of computer is not used in computer graphics. Its primary use is to solve differential equations for research. *Digital* computers use numbers for manipulation of data, and are the most common type used today. When the word computer is used in this text, it refers to digital computers.

Information put into a digital computer is converted into numbers based on the binary number system. The *binary number system* is a base two number system using ones and zeros. Series of ones and zeros are used to represent information processed by the computer. For example, the number 2 is represented in binary as 10, and the number 10 is represented in binary as 1010. Although the need to know the binary number system is not important to a CAD operator, it does help the operator to better understand how a computer works, and why the computer is called a digital computer.

The First Electronic Digital Computer

The first large electronic digital computer in the United States was developed at the University of Pennsylvania in 1946 and was called the ENIAC (electronic numerical integrator and calculator). The computer was made up of 18,000 vacuum tubes, 70,000 resistors, 10,000 capacitors, and 6,000 switches and it occupied the space of a two-car garage. The computer was reprogrammed by rewiring the circuits. It could perform 5,000 additions and subtractions per second. Today, most home computers can do the same number of calculations and fit easily onto the top of a desk.

ENIAC is an example of the first generation of computers. Looking back into history, one can find that computers can be grouped into four generations of computers based on the electronic technology of the time.

FIGURE 2-1 The Mark I automatic calculator used mechanical and electrical components instead of electronic. *(Courtesy of the International Business Machines Corporation)*

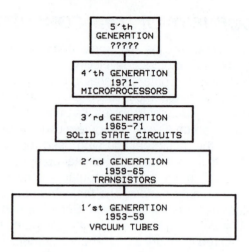

FIGURE 2-2 The four generations of computers. Each succeeding generation has become smaller, more powerful, less costly to purchase and operate, and has more memory and increased computing speeds.

The first generation lasted until approximately 1959 and was characterized by the use of vacuum tubes. They were large, bulky, expensive to buy and power, and often unreliable. Instructions and data were entered through the use of punched cards. The second generation ran from about 1959 to 1965 when vacuum tubes were replaced with the transistor, which reduced the cost, energy requirements, and heat generated by the computer. They were also more reliable with increased storage capacity. The third generation from 1965 to 1971 saw the use of solid state circuitry using Silicon Chips for integrated curcuits. The fourth generation of computers started in 1971 and continues today. The fourth generation is characterized by the use of microprocessors (chips). A *microprocessor* is a system of miniaturized circuits necessary to process a program. These silicon chips are smaller than a dime and have thousands of electronic circuits etched onto a single chip. These chips are then mounted or packaged into a unit about the size of the return key on a typewriter. This is the reason that the home computer can do the work of ENIAC, only in a much smaller package. The circuitry that made up ENIAC has virtually been reduced to a handful of chips in the last thirty years.

Of course, the dates mentioned vary somewhat because of overlapping technology, but they are an indication of the rapid development and improvement in the computer. With each generation there was an increase in computing speeds and memory capacity, and a decrease in memory cost and the cost of computers. The size of the computer kept shrinking, and environmental restrictions became less severe with fewer hardware breakdowns occurring.

THE DEVELOPMENT OF COMPUTER GRAPHICS

The use of computer graphics can trace its beginnings back to 1952. The Whirlwind computer at Massachusetts Institute of Technology

(MIT) was installed that year to draw simple pictures. The SAGE air defense command system was installed in the mid 1950s and used light pens to identify targets on a display screen. The use of computers to display graphics was not to become practical until the 1960s.

In 1962, Dr. Ivan Sutherland published his doctoral thesis, entitled "Sketchpad: A man machine graphical communication system." This one event probably did more to promote interactive computer graphics than any other single event. The word *interactive* in graphics means the need for human intervention in the operation of a computer to complete a task. Sutherland's thesis proved to many that interactive computer graphics could become a viable and productive method of using computers for graphical display. Soon afterward, research projects were initiated by MIT, Lockheed Aircraft, Bell Laboratories, and General Motors Corporation. It was at this time that the enormous potential for computer-assisted drafting became clear to manufacturers in the automobile and aerospace industries. This research was to be the catalyst for the commercial success of computer graphics and CAD in the 1970s.

The '70s saw a tremendous advance in hardware technology and the commercial introduction of CAD systems and related hardware from such companies as Applicon, Calma, Computervision, Houston Instruments, and Summagraphics. Advances in microprocessor technology and increased sales led to a growing number of graphic suppliers entering the market. In the early '80s, there were more than 100 suppliers of CAD systems and software, and a number of organizations sponsored CAD/CAM conventions and conferences as a means of sharing information. CAD/CAM revenues topped $1 billion in the early '80s as the technology improved and prices continued to fall. After a slow start, computer graphics grew dramatically, and the outlook is that it will continue to grow at a rapid rate.

WHY THE SUDDEN GROWTH IN COMPUTER GRAPHICS?

Since the late '70s there has been a dramatic increase in the num-

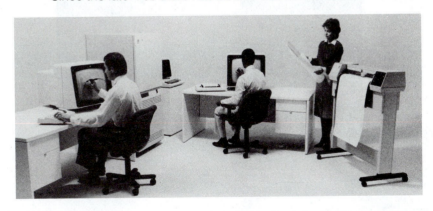

FIGURE 2-3 The rapid development of computer graphics has led to the use of powerful CAD workstations in drafting-design. (*Courtesy of the International Business Machines Corporation*)

ber of CAD systems on the market and in the number of industries using them. No one event produced this increase in CAD, but there are a number of important reasons. Contributing to the increased use of CAD by industry are the rapid development in the microcomputer due to improved microprocessor technology, the dropping cost of memory, and the increased number of vendors supplying CAD. Another major reason for the growth in CAD is competition among rival companies both in the United States and abroad. Industries are finding that CAD must be used in order to remain competitive in such fields as electronics. The decrease in turnaround time in design and increases in productivity are two ways that CAD can make a company more competitive. CAD is and will continue to be the most productive method for drafter-designers to perform their job.

THE FUTURE OF COMPUTER GRAPHICS

A growth rate of approximately 15% is expected in computer graphics through the 1980s and early 1990s. This multibillion dollar industry will generate fierce competition among the manufacturers in the next few years. This competition will produce some changes in the methods and hardware used to create graphics. Of course, some advances in computer technology will also cause many changes in computer graphics.

One of the expected changes will be in the method of displaying graphics. The 1990s will see the increasing use of flat plasma display devices and liquid crystal display screens for computer graphics. Present displays are too bulky for some uses. Flat screen technology can promise lighter weight, less bulk, and lower power consumption. Figure 2-4 shows a flat plasma display device. Color display devices are now the preferred method for displaying graphics.

FIGURE 2-4 The IBM 581 flat plasma display device. *(Courtesy of the International Business Machines Corporation)*

FIGURE 2-5 Drawing being digitized by hand. With improved technology and software developments, digitizing will be done automatically by computers. *(Courtesy of Bausch & Lomb)*

As semiconductor technology improves, computing costs will fall dramatically as speed is increased. There will also be a drop in the cost of computer hardware. Hardware performance and reliability will improve over today's standards.

Voice-controlled terminals will become more common in the coming years. Voice-controlled terminals will use the operator's voice to activate commands. The operator's voice will become the primary method of interacting with the computer instead of a mouse or other input device.

As software becomes more sophisticated, CAD will become more efficient and software packages, such as 3-D modeling, will become more common in even the lower-priced CAD systems. Another software development will be for the digitizing of drawings. Drawings produced by traditional methods can be entered into a CAD system by a process called digitizing. *Digitizing* converts graphic data (drawings) into digital data that can be used in the CAD system. At present, most digitizing is done by hand, as shown in Figure 2-5. However, the software and technology has been developed that will enable computers to recognize patterns, thus allowing the automatic digitizing of drawings.

THE FIFTH-GENERATION COMPUTER

Of all the improvements in computer graphics expected in the coming years, the development of a fifth generation of computers may have the greatest impact on computer graphics and the computer as we know them today.

Japan is undertaking the development of a fifth generation of computers, with a prototype ready by 1990 as its goal. This new generation of computers will have artificial intelligence as one of its features. The com-

FIGURE 2-6 Traditional tools of the trade are being replaced by a new tool: CAD. *(Photo by Arthur A. Molitierno)*

puter will be able to make inferences as a human does, and to solve problems that it has not directly been asked to solve. It will also be able to retrieve data from a larger data base and solve problems in a manner similar to that of the human brain. This computer of the future will be much easier to use than those of today because of voice and pattern recognition capabilities instead of keyboard input. Programming will become much easier, the computer will be more compact, and it will be less costly to operate.

The effects of this very ambitious project on the future of computer graphics is difficult to forecast. Whatever the outcome of this project, computer graphics will continue to change, evolving and growing into a major industry. These changes will have a direct effect on CAD over the coming years, making the drafting-design field a very dynamic and challenging occupation.

Chapter Two GLOSSARY

Analog computer—uses variations in physical quantities, such as electrical voltage to control or solve problems.

Binary number system—a base two number system using ones and zeros to represent information processed by a digital computer.

Digital computer—a device that uses numbers for the manipulation of data.

Digitizing—the process of identifying, locating or selecting a menu item, entity or point through an input device.

Interactive—refers to the need for human intervention in the operation of a computer in order to complete a task.

Microprocessor—the miniaturized electronic circuits necessary to process a program; the "brains" of a computer.

Chapter Two REVIEW

1. List the generations of computers used to group computer technological advances.
2. Describe some of the reasons for the sudden growth of computer graphics.
3. Explain some of the future developments expected in computer graphics.
4. What does interactive computer graphics mean?
5. Describe a microprocessor and its function in a computer.
6. For what is digitizing used?

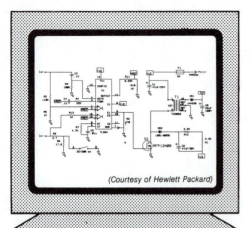
(Courtesy of Hewlett Packard)

Chapter Three

The Components and Operation of a Computer

To the drafter, the "computer" in CAD may seem to be a confusing mass of circuits and hardware. However, the computer is not quite so confusing if the drafter understands its basic operation and is familiar with its components. As with any technology, a jargon is associated with computers. Your chances of mastering the subject will increase if you understand the terms used with computers. This chapter presents the fundamentals for learning the language associated with computers. Chapters Four and Five build upon this foundation to increase your knowledge of the computer terms essential in becoming a CAD operator. You will then be capable of transferring your familiarity and experience with one CAD system to any other system.

Not all CAD systems are alike in the method of producing a drawing. The system on which you learn may not be the same type that you will have on the job. How then can CAD training help if you are not trained on the same system that you are going to work with? One reason is that all CAD systems have certain characteristics that are similar, such as a keyboard for input. Another reason is that CAD training allows a drafter to become comfortable in using computers. Many of the skills learned on a CAD system are easily transferred to different systems. Think of it as driver's training. When learning to drive, you may have been taught with one make of car. But, you probably had little trouble transferring the skills learned with that car in order to drive another make. All cars have certain parts that are similar, such as a gas pedal, brake pedal, steering wheel, and directional indicators. Transferring skills from one CAD system to another is similar to the ease in making the transition from one car model to another.

The next three chapters explain the operation of the computer and point out the similarities and differences among CAD systems. This chapter introduces you to the five components of the basic computer. Learning about these five components will help the drafter to understand how a computer operates and some of the terminology associated with computers.

Included in this chapter is an explanation of the three classifications of computers. Also discussed are the languages used to communicate with a computer.

OBJECTIVES

After completing this chapter, you will be able to

- identify and describe the five components of a computer.
- define the *central processing unit* (CPU) and identify its components.
- name the three classes of computers.
- define computer *language* and name the common types associated with CAD.
- define the terms *RAM* and *ROM*.
- list the common types of memory storage devices.
- define the terms *bit* and *byte*.
- define the units of measure for memory storage.

THE FIVE COMPONENTS OF A COMPUTER

The *computer* is a device or tool used to process *data* consisting of input, output, memory and the central processing unit (CPU). For the drafter, the *computer* is a tool used to automate drafting tasks. *Data* are raw facts and figures represented by such symbols as letters, numbers, or special symbols. Separating the computer into its important parts will help the drafter to understand the operation and terminology associated with computers. The five main components of a computer are:

1. *Arithmetic logic unit (ALU)*—circuitry that performs the logic and mathematical operations associated with a computer, such as addition, and control of data.
2. *Memory*—circuitry that stores data input or processed by the computer. This memory can be of two forms: random access memory (RAM) or read-only memory (ROM).
3. *Input devices*—mechanism used to interact with the computer.
4. *Output devices*—used to display data in a form understandable to the operator.
5. *Controller*—circuitry used to regulate all operations taking place in the computer.

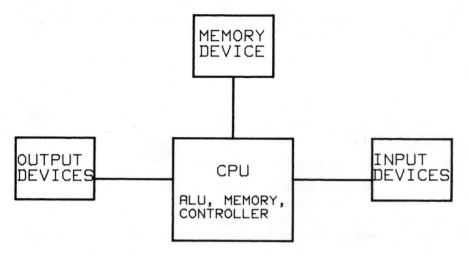

FIGURE 3-1 The main components of a typical computer.

Central Processing Unit

In the past, the five components that make up a computer had been five distinct units. Now, because of microminiaturization of circuits, the ALU, controller and part of the memory of a computer can be combined into one package called the *central processing unit* (CPU). The CPU is considered to be the "brains" of the computer because it controls all the operations of the computer. It is sometimes referred to as a *microprocessor* in smaller computers because all the components of a CPU can fit on a single integrated circuit (IC) which is often referred to as a chip. The word *chip* is often used in place of *integrated circuit*.

The CPU, an integrated circuit (IC) chip, is one of the miracles of modern electronic technology. The hundreds of circuits, resistors, gates, and other electronic components necessary in a computer have been reduced and etched onto a silicon wafer or chip not much larger than the head of a pin. Figure 3-2 shows an integrated circuit (IC) chip. Once the chip has

FIGURE 3-2 The integrated circuit (IC) is the square located on the petals of the flower. Present technology allows thousands of circuits to be etched onto these small chips. *(Courtesy of Intel Corp.)*

FIGURE 3-3 A packaged integrated circuit (IC) with the top layer removed revealing the chip. *(Courtesy of Intel Corp.)*

been produced, it is packaged into a larger unit such as the one shown in Figure 3-3. After packaging, the chip is still referred to as a chip.

Ironically, the design of the chips used in today's computers would not have been possible without CAD. The circuits used in chips are so small that they are designed using CAD. Figure 3-4 shows a drawing of a circuit designed on a CAD system. This is an excellent example of how one technology is dependent upon another in order to advance.

Microprocessors work only with *binary numbers* (ones and zeros). All instructions and numbers required for an operation must first be converted into groups of binary numbers before the CPU can handle them. The number of binary digits that a CPU can handle determines the classification used to group microprocessors. If the CPU can handle groups of eight binary digits, it is called an 8-bit microprocessor. A *bit* is a binary digit that can have only one value: either a one (1) or a zero (Ø). A 16-bit microprocessor can handle sixteen bits at a time, a 32-bit microprocessor can handle thirty-two bits at a time, and so on. Usually, the larger the groups of digits (called *word size*) that a CPU can handle determines the speed at which the computer performs a task.

Most CAD systems have 16- or 32-bit microprocessors. The type of microprocessor in a CAD system is important because it usually determines the speed of executing a drawing. The faster the CAD system produces a drawing, the more efficient the operator becomes. Knowing the type of microprocessor that a CAD system contains is important when evaluating a system for purchase. It should be noted that good software or logic circuitry with an 8-bit processor could outperform or be faster than a 16-bit processor with poor software or logic circuitry.

The memory section contained in the CPU is used to store results of calculations or program instructions, and is usually called *main memory*. Outside of the CPU, there usually are more memory chips used to store additional information, such as drawings. This type of memory is called *random access memory* (RAM). RAM memory is volatile (transitory); it is memory that will be lost forever when the computer is turned off. *Read-only memory* (ROM) is another type of computer memory circuit. ROM is instructions

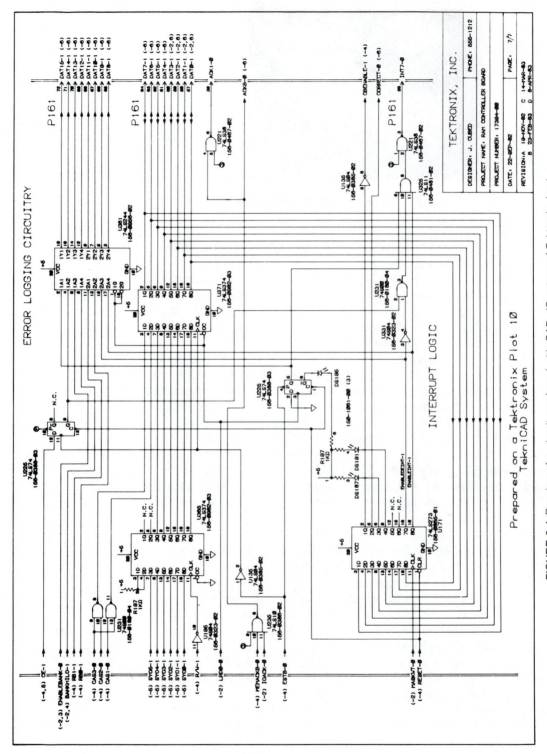

FIGURE 3-4 Drawing of a circuit produced with CAD. (Courtesy of Tektronix, Inc.)

that have been permanently programmed during the manufacture of the chip. This memory is permanent and will not be erased when the computer is turned off.

Permanent Memory

Obviously, using computers for drafting would not be a very efficient method to produce drawings if all the work were lost when the computer was turned off. Logically, there are methods of permanently storing drawings, software, or any information processed by a computer. Common methods of permanently storing information are shown in Figure 3-5. These methods include magnetic tape, 5-½" or 8" floppy disk and 3-½" microdisks, cassette tape, and punched tape.

Magnetic tape is most commonly used on very large computers or it can be used in CAD to archive drawings. *Archiving* is used to permanently store a drawing or data on magnetic tape which is then stored in a safe place to prevent accidental loss due to fire or some other unexpected event.

Floppy disk and hard disk storage are the most common methods of storing information on CAD systems. Floppy disks are available in different sizes, and they can store in excess of 1 million bytes of information. Hard disks usually come in an airtight fixed cabinet, commonly called a *Winchester disk drive,* or in cartridge form, sometimes referred to as a *disk pack.* Storage capacities of hard disks greatly exceed those of floppy disks with Winchester technology achieving the greatest storage densities in the range of 5 to 100 million bytes of information. Punched cards are not in great use today; however, the use of punched tape for memory storage is still a common method used to store programs for computer-controlled machine tools.

Taking the Byte out of Bits

Memory devices come in various sizes and storage capacity. The storage capacity of a memory device is measured in bytes. A *byte* is a group

FIGURE 3-5 Various media used for the permanent storage of data processed on computers.

of eight bits. Remember that a bit is a single binary digit having a value of 1 or Ø. So, two bytes contain sixteen bits, three bytes contain twenty-four bits, and so on. Because the memory capacity of modern computer equipment is so great it is usually measured in terms of kilobytes (K) or K-bytes. A K of memory is equivalent to 1024 bytes of memory. In general terms, a *kilobyte* of memory is usually referred to as 1000 bytes of information. If the storage capacity of a floppy disk is 500 K, it has a storage capacity of approximately 500,000 bytes of information. Most microcomputers today require 512 K or 640 K of RAM memory to use CAD software.

Memory device technology is advancing at a rapid rate and a kilobyte of memory is becoming too small a unit of measure for the storage capacity of some devices. Technology dictates that every several years the amount of data stored in a given medium will double. Some storage devices have capacities in the millions of bytes. When this occurs, memory is measured in units of megabytes (Mb). A *megabyte* of memory is 1 million bytes or 1000 K-bytes of information. It is not uncommon for a hard disk drive to have a storage capacity of ten Mb or 10 million bytes.

Input Devices

Input devices are one of the five components that make up a computer. The most common method of entering information in a computer is with the *keyboard*. Using the keyboard as the input device on a CAD system would be a very inefficient method of making a drawing. Thus, alternative input devices have been developed for CAD, such as *mice* and *tablets*. These input devices used for CAD will be covered in greater detail in Chapter Four. Figure 3-6 shows some common input devices used with computers.

FIGURE 3-6 CAD workstation with two input devices, control dials, and joystick. *(Courtesy of Vector General)*

FIGURE 3-7 Microcomputer-based CAD workstation. *(Courtesy of VersaCAD)*

Output Devices

The most common output device on a computer is the monitor or cathode ray tube (CRT). A *cathode ray tube* is very similar to the television screen found in a home, and it is used to display information processed by the computer or input by the operator. Another common output device is a *printer.* In a CAD system, another output device is the *plotter*, which is used to make a hard copy of the drawing created on a graphics system. Figure 3-8 shows a plotter used to print a drawing made on a CAD system. Output can also be sent directly to a numerical control (NC) machine tool.

THE CLASSIFICATION OF COMPUTERS

In the past, a computer could be easily classified according to its size, speed of operation, and the degree of difficulty of programs that it could execute. Today, because of the microminiaturization of circuits and the similarities among capabilities, classifying computers is becoming more difficult. However, computers can generally be classified into three categories: mainframe, minicomputer, and microcomputer. These classifications are determined by the type of CPU used.

Historically, a mainframe computer used to handle more devices, and had faster speed, more memory, and possibly multiple users, while a minicomputer (mini) had much less capability. Today, although the mainframe and mini terms are still in use, the distinction is not always clear. Now

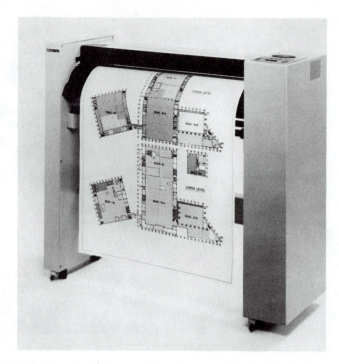

FIGURE 3-8 A large-format multicolor pen plotter. *(Courtesy of Hewlett-Packard)*

minis outperform the mainframes of yesterday by having larger memory and faster speeds, and by handling more users. Even the microcomputers (micros) are using 32-bit word size and performing much like the mini and small mainframe. In the future, the distinction among all classes of computers may become even less than it is today.

Mainframe Computer

A *mainframe* computer is a very large computer with the CPU housed in a remote, environmentally controlled setting. This environment is usually a room that is air conditioned and relatively free from dust. The CRTs or workstations are located at some distance from the CPU. These types of computers are the most powerful of the three, with very rapid computing speeds, large amounts of memory, and the ability to process even the most difficult of programs. Very powerful CAD/CAM software programs can run on this hardware with multiple engineering workstations.

Minicomputer

The second class of computer is the minicomputer. In the past, the distinction between the mainframe computer and this computer was much greater. *Minicomputers* are usually housed in a controlled environment similar to mainframe computers. The difference lies in the fact that minicomputers usually operate at slightly slower speeds and are not able to process

FIGURE 3-9 A large 32-bit minicomputer used for CAD. *(Courtesy of Intergraph Corp.)*

some of the more difficult programs that can be handled by the mainframe computer. Figure 3-9 shows a super-minicomputer. Many types of CAD/CAM software programs run on minicomputers.

Microcomputer

The microcomputer is the newest addition to the classification of computers. A *microcomputer* is basically a processor on a chip, and its origin can be traced to the development of the microprocessor in the mid '70s. Typically, microcomputers run slower, cannot handle some more difficult programming operations, and do not need as strict environmental controls as do mainframe computers and minicomputers. However, the development of new classes of microprocessors has allowed microcomputers to develop more power and speed. With the addition of a co-processor chip, most currently-used microprocessors can reduce processing time by one third. A recent development has been the first microcomputer using the 32-bit microprocessor chip; this, in turn, will allow further development of CAD software capabilities and power. Figure 3-10 shows a number of microcomputers used to display graphics.

There are well-written CAD software programs for all classifications of computers. The major differences among the classifications of computers using CAD are the speed of executing a drawing function, such as cross-hatching; the ability to share a common data base between terminals; and the speed of executing more difficult engineering types of programs, such

FIGURE 3-10 8- and 16-bit microcomputers used to display graphics. *(Courtesy of Hewlett-Packard)*

FIGURE 3-11 The IBM 4331 computer and components. On the left are magnetic tape drives at the rear are hard disk drives, at the far right is a printer, and in the center front is the CRT workstation. *(Courtesy of the International Business Machines Corporation)*

as 3-D modeling. With well-written CAD programs, computers will make the drafter a much more productive person, regardless of the type of CPU on which the software is being run.

COMMUNICATING WITH COMPUTERS

Computers are certainly marvelous devices that have proved to be useful for many tasks, such as drafting, but it is necessary to know how to communicate with a computer in order for it to perform these useful tasks. A computer only understands a certain language presented to it in a specific manner. The set of instructions necessary to command a computer to perform a specific task is called a *program*. These instructions must be written in a language that can be understood by the computer. A *language*, such as BASIC, is a written set of words used to communicate with a computer. BASIC is one type of high-level language that is composed of English words, statements, and syntax. There are more than 200 languages for communicating with computers. Of these 200, FORTRAN, PASCAL, and BASIC are the most common languages used in CAD. Figure 3-12 shows three short programs written in different languages that will accomplish the same task. The type of language used in programming depends upon the language used by the specific computer hardware.

A person who writes programs for computers is known as a *programmer*. A drafter does not need to be a programmer to become a CAD operator. Programmers write software that allows the drafter to communicate with the computer in order to produce drawings. However, a drafter may find it beneficial to learn the basics of programming. Doing so will make the CAD operator a more useful employee because of the opportunities for writing parametric programs, and for the occasional need to write a short

BASIC PROGRAM

```
5 REM    THIS IS A WMBASIC EXAMPLE PROGRAM.
10 REM   THIS PROGRAM ESTIMATES THE AREA UNDER A CURVE USING
15 REM   RECTANGULAR APPROXIMATION
20       let A = 1
30       let B = 2
40       let N = 100
50       let DX = (B - A)/N
60       let X = A
70       let T = 0
80       for I = 1 to N
90            let X = X + DX
100           let T = T + X * X
110      next I
120      A1 = DX * T
130      print " FOR THE CURVE Y = X SQUARED "
140      print " FROM X VALUES ";  A; " TO "; B
150      print "THE APPROXIMATE AREA UNDER "
160      print " THE CURVE AND ABOVE THE X-AXIS IS ";A1
170      end
```

FIGURE 3-12 Examples of computer programs that perform the same task written in versions of BASIC, FORTRAN, and PASCAL.

```
C        THIS IS A FORTRAN EXAMPLE PROGRAM.
C        THIS PROGRAM ESTIMATES THE AREA UNDER A CURVE USING
C        RECTANGULAR APPROXIMATION.
C
         INTEGER N, I
         REAL A, B, DELTAX, TOTAL, REALN, X, AREA
         A = 1.
         B = 2.
         N = 100
         REALN = N
         DELTAX = (B - A)/REALN
         X = A
         TOTAL = 0
         DO 10  I = 1, N
                 X = X + DELTAX
                 TOTAL = TOTAL + X * X
10       CONTINUE
         AREA = DELTAX * TOTAL
         PRINT *, ' FOR THE CURVE Y = X SQUARED '
         PRINT *, ' FROM X VALUES ', A, ' TO ', B
         PRINT *, ' THE APPROXIMATE AREA UNDER '
         PRINT *, ' THE CURVE AND ABOVE THE X-AXIS IS ', AREA
         STOP
         END
```

PASCAL PROGRAM

```
{THIS IS A PASCAL EXAMPLE PROGRAM.}
{THIS PROGRAM ESTIMATES THE AREA UNDER A CURVE USING}
{RECTANGULAR APPROXIMATION}
VAR
         A, B, DELTAX, TOTAL, REALN, X, AREA: REAL;
         N, I: INTEGER:
BEGIN
         A := 1.0 ;
         B := 2.0 ;
         N := 100 ;
         REALN := N ;
         DELTAX := (B - A)/REALN ;
         X := A ;
         TOTAL := 0 ;
         FOR I := 1 TO N DO
         BEGIN
                 X := X + DELTAX ;
                 TOTAL ;= TOTAL + X * X
         END ;
         AREA := DELTAX * TOTAL ;
         WRITELN( ' FOR THE CURVE Y = X SQUARED ') ;
         WRITELN( ' FROM X VALUES ',A , ' TO ', B ) ;
         WRITELN( ' THE APPROXIMATE AREA UNDER ') ;
         WRITELN( ' THE CURVE AND ABOVE THE X-AXIS IS ', AREA)
END.
```

The codes used in these programs are for illustration purposes only.
They were run on a Digital Equipment Corporation VAX 11-750 CPU.
The versions of the programming language used here may vary slightly
on other computers.

FIGURE 3-12 (continued)

program that may be useful in the drafting office. With 70,000 lines of code not uncommon for programs written for CAD, this type of project is better left in the hands of trained computer programmers. The drafter's skill lies in the efficient and rapid execution of the CAD program to produce drawings.

CAD APPLICATION PROGRAMS

Programmers can write software for almost any desired task. Throughout the years, programs have been written for many applications related to drafting-design and manufacturing. Programs written for a specific topic, such as architectural drawing or word processing, are called *application programs*. The following is a listing of some of the application programs created for CAD.

- Finite element analysis
- 3-D modeling
- Cartography
- Marine
- Aeronautical
- Petrochemical
- Nuclear
- Numerical control
- Microprocessor design
- Printed circuit boards
- Mechanical analysis (stress, strain)
- Bill of materials
- Business graphics, charts and graphs
- Technical illustration
- Heating, vents, air conditioning (HVAC)
- Fluid mechanics
- Mechanical drawing
- Architectural drawing
- Electrical
- Electronic schematic
- Piping
- Civil
- Plant layout, space planning

Of course, application software packages are available other than those listed. As is evident, there is seemingly a software package available to assist the drafter in any drawing project.

Chapter Three GLOSSARY

ALU—arithmetic logic unit is circuitry that performs the logic and mathematical operations associated with a computer, such as addition and control of data.

Application program—a computer program written for a specific topic, such as architectural drawing or word processing.

Archive—a process that stores data on magnetic tape which is then stored in a safe place to prevent accidental loss.

Bit—a single digit of the binary system having a value of one or zero.

Byte—a group of eight bits.

Chip—an integrated circuit etched onto a silicon wafer.

Computer—a device or tool used to process data consisting of input, output, memory, and a central processing unit (CPU).

Controller—circuitry used to regulate all operations that take place in the computer. One of the components in a central processing unit (CPU).

CPU—central processing unit is an integrated circuit (IC) containing the ALU, controller, and part of memory. It is considered to be the "brains" of the computer, and is sometimes referred to as a *microprocessor*.

CRT—cathode ray tube or monitor is an output device, similar to a television screen that is used to display information processed by the computer or entered by the operator.

Data—raw facts and figures represented by such symbols as letters, numbers, or special symbols.

High level language—standardized English like programming languages such as BASIC or Fortran.

IC—integrated circuit is a microminiaturized circuit etched onto a silicon wafer sometimes referred to as a *chip*. The CPU and circuits used for random access memory (RAM) and read-only memory (ROM) are examples of ICs.

Kilobyte (K)—a unit of measure for memory storage equivalent to approximately 1000 bytes or 1024 to be exact.

Language—a written set of words or symbols used to communicate with a computer. Languages are identified as BASIC, FORTRAN, PASCAL, COBOL, and others.

Mainframe computer—a large computer housed in a remote, environmentally controlled setting. Typically, the most powerful type of computer.

Megabyte (Mb)—a unit of memory storage equivalent to approximately 1 million bytes.

Memory—circuitry (random access memory or read-only memory), or media (floppy disk, magnetic tape, and so forth) that stores processed data inside of the computer.

Microcomputer—a computer based upon the microprocessor. Strict environmental controls are not needed.

Minicomputer—a computer whose processing speeds range between those of the microcomputer and the mainframe computer. Strict environmental controls are needed.

Operator—a person who is trained to use an application software package on a computer.

Program—the set of instructions, arranged in a logical sequence, used to command a computer to perform a specific task.

Programmer—a person trained to write programs to control the operation of a computer.

Random access memory (RAM)—volatile (transitory) memory used to store data temporarily during processing.

Read-only memory (ROM)—computer instructions that were permanently programmed on a chip during its manufacture.

Chapter Three REVIEW

1. List the five components that make up a computer.
2. What components of a computer make up a CPU?
3. Name the three classes of computers.
4. Explain the difference between an operator and a programmer.
5. Is a drafter trained in CAD considered to be a programmer or an operator?
6. What is the function of a CPU?
7. List some of the methods used to permanently store computer memory.
8. Relate these memory storage units: bit, byte, K-byte, megabyte.
9. List examples of input devices used on computers.
10. List examples of output devices used on computers.
11. List some example application programs related to CAD.
12. What is a computer language and which types are most commonly used in CAD?
13. Define the term *computer*.

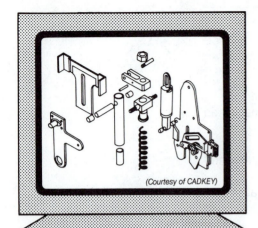

(Courtesy of CADKEY)

SECTION 2

TYPICAL CAD SYSTEM

Chapter Four

The Components of a CAD System

This chapter introduces the reader to the components and peripherals that make up a CAD system. A *peripheral* can be described as auxiliary equipment or a device connected to the CPU. This chapter describes the various peripheral devices used in conjunction with CAD. With more than 100 CAD/CAM systems on the market, peripherals used in a CAD system come in many different forms.

For a variety of reasons, it is important that a trained CAD operator be familiar with the numerous hardware devices used in different CAD systems. To be truly flexible after being trained on one CAD system, the CAD operator should then become familiar with the other forms of devices found in different CAD systems. If you are trained on a system that uses a light pen for input, but on the job you operate a CAD system that uses a puck for input, you should feel confident enough so that learning to use a CAD system with a different input device will not be a problem for you. Being able to describe these peripheral devices will also prepare you to evaluate CAD systems for purchase, because some of the advantages and disadvantages of each device are also discussed in this chapter. Also described are the types of software used in CAD, and the user associated with CAD.

OBJECTIVES

After completing this chapter, you will be able to

- list the components of a CAD system.
- identify the different types of CRTs used with CAD.
- name the various input devices used with CAD.
- name the various output devices used with CAD.
- describe the different types of software used with CAD.
- discuss the importance of users associated with CAD.
- identify the types of memory devices used with CAD.
- identify the hardware components in the CAD system that you are using.

THE COMPONENTS OF A CAD SYSTEM

For a computer to be considered a CAD system, it must have certain components that will enable it to draw faster than using traditional tools. These components include the hardware, the software, and the human operator, Figure 4-1. The hardware and software devices used with CAD are varied; for this reason a number of different subcategories are described under the listing of each component.

HARDWARE

The hardware used in CAD can be subdivided into five categories: the CPU, display devices, input devices, output devices, and memory devices. Many CAD systems have more than one hardware device in each category. For example, Figure 4-2 shows a CAD/CAM system that has three different input devices: the keyboard, the joystick, and a function board. So, keep in mind that a CAD system can have more than one input or output device available for the operator to use. In fact, it is usually an advantage to have more than one device available to the CAD operator, as this makes the system more flexible and allows the operator to be more efficient.

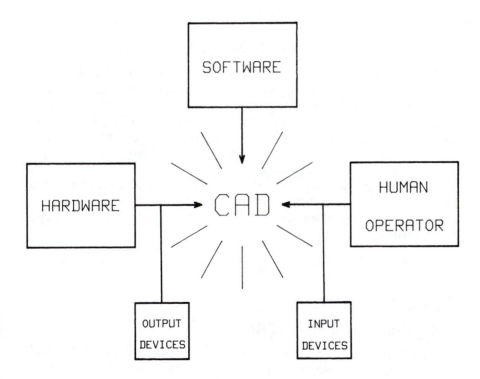

FIGURE 4-1 The three components of a CAD system.

FIGURE 4-2 CAD/CAM system having three input hardware devices. *(Courtesy of McDonnell Douglas Automation Co.)*

THE CPU (CENTRAL PROCESSING UNIT)

To the drafter, the CPU is a transparent device that no longer concerns the drafter once drawing begins. The only time that a drafter is concerned with the CPU is when evaluating CAD systems for purchase or when the CPU is not functioning properly. A CAD system cannot function without a CPU and, therefore, this component is said to be the "brains" of the computer and controls its operation. A quick review of Chapter Three will give you a more detailed description of a CPU.

A CPU is just a small electronic device usually known as a chip. But, many times the keyboard and housing for the CPU and its auxiliary circuits is referred to as the CPU. Figure 4-3 shows a standalone CAD system and CPU.

DISPLAY DEVICES

The Cathode Ray Tube (CRT)

The CRT is one of the most important peripherals on a CAD system. A CRT can be compared to a television screen. The type of CRT determines the quality of the drawing produced on the screen and the ease in making changes to that drawing on screen. Looking at a CRT for up to eight hours a

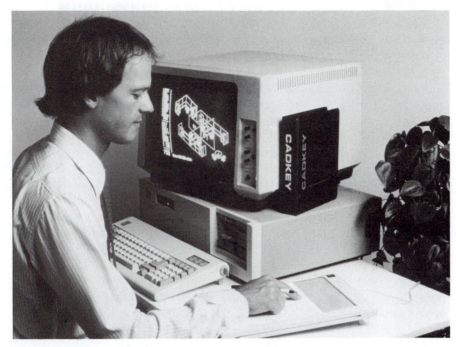

FIGURE 4-3 Microcomputer-based CAD system with the CPU and support circuitry housed in the cabinet below the display monitor. *(Courtesy of CADKEY)*

day can be a great strain on the operator. So, the type of CRT used on a CAD system can have an effect on the performance of the designer. Some concern has been expressed that looking at a CRT for extended periods of time may be harmful to the operator's eyesight. However, little proof to defend this claim has been found, but research continues.

The most common types of CRTs used in CAD are the raster display, vector storage and vector refresh. These three types are described next, as well as two types that will become more common in the years ahead: the plasma display, briefly described in Chapter Two; and the liquid crystal display.

The Raster Display Terminal

The raster display terminal is the most common method of displaying graphics in CAD. A raster display can be compared to the picture tube in a television. Take a close look at a television screen and you will see that it is made up of thousands of small dots. These dots are called *pixels,* which is short for picture elements. Pixels are arranged horizontally and vertically on the screen as shown in Figure 4-4. The number of pixels that fill the screen determines the resolution of the screen. The smaller or denser the number of pixels the better the resolution. The *resolution* of a screen can be described as the clearness or sharpness of the display. Figure 4-4 shows how each individual pixel would be turned on to draw the numeral 8 on a raster display.

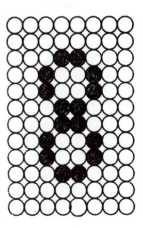

FIGURE 4-4 Enlargement of a raster display device showing how the individual pixels are lit to form numbers, letters or lines.

A raster display is usually classified according to the resolution or pixels that make up the screen. For example, a CRT with a resolution of 680 by 520 means that there are 680 pixels horizontally across the screen and 520 pixels vertically on the screen. This display would have 353,600 pixels or dots on the screen. Each one of these dots must have a memory location in the computer. As the cost of memory falls, the sharpness of raster displays will increase as more pixels can be used economically for display. Displays offering resolution sharper than a 35-mm photograph will not be uncommon in the future. In color displays, each dot can be assigned a color.

The main advantages of the raster display over other types is its relatively low cost, and the speed with which it redraws an image. For these reasons, it is the most popular type of display used in CAD. However, raster displays do have one distinct disadvantage as compared to other types of display devices. This shortcoming becomes apparent when drawing images that are not exactly horizontal or vertical. Figure 4-5 shows an example of this phenomenon known as the *"jaggies," "aliasing,"* or *"stair-stepping."* Because the image produced must follow the lines of dots or pixels that are lined up horizontally and vertically, a line that is drawn at some angle to these rows of dots will end up being out of alignment. See Figure 4-6. This stair-stepping becomes less pronounced as the resolution of the display increases.

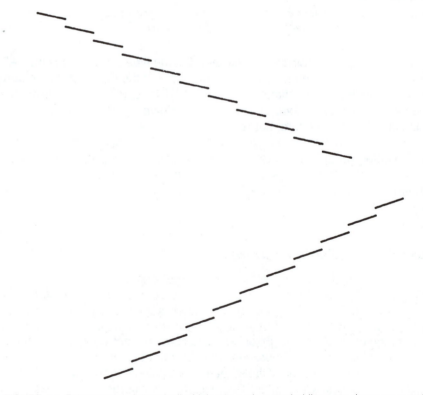

FIGURE 4-5 Enlargement of "jaggies" which occur when angled lines are drawn on a raster display.

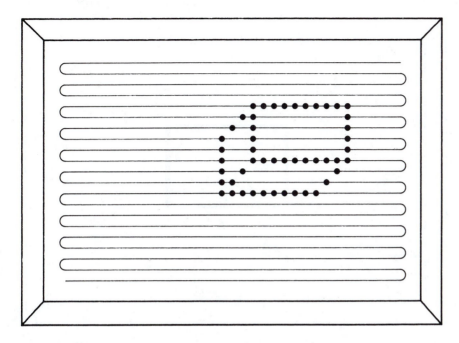

FIGURE 4-6 Enlargement of a moving electron beam on a raster screen as it scans back and forth across the screen. The pixels are lit by the beam forming lines or patterns on screen.

In the future, this disadvantage should become less of a factor as screen resolutions increase. However, if a high-resolution screen is needed for design, raster displays may not be the best type of display device to use. For this reason, two other display devices may be used in CAD.

Vector Storage Displays

One alternative display device that can be used is the vector storage or direct-view storage tube (DVST). The vector storage device will maintain its image for an extended period of time. To draw any line at any angle, the end points are located and the screen phosphors are turned on, resulting in a clean, straight line or vector. See Figure 4-7.

One of the greatest advantages of the vector storage display is that it does not suffer from the jaggies as does the raster display. This is a very important factor in some design work when the image is complicated, such as a PC board design. However, there are some distinct disadvantages to such devices. Images can be moved or deleted only by deleting the whole image from the screen and redrawing or refreshing the screen. A highly complex drawing can take time to redraw entirely, so this type of display is a drawback to a designer who makes frequent deletions or changes. Also, when the image is deleted, a bright flash is produced which some operators may find to be annoying. Another possible drawback is that these displays are easier to see when used in low lighting situations.

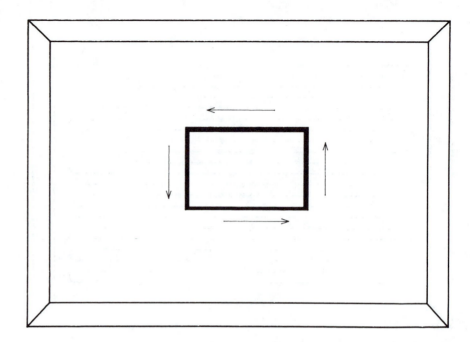

FIGURE 4-7 A direct-view storage tube display draws lines instead of lighting pixels, allowing for much better resolution. Once an image is on screen, it remains until the whole image is erased.

Some DVST terminals have a color-enhanced refresh feature. This recent development allows a repaint feature that permits recently produced images to be selectively deleted without total picture erasure. This technology helps to overcome some of the disadvantages of DVST when compared to raster displays.

Vector Refresh Displays

The vector refresh display combines features of the raster and vector storage CRTs. This device is sometimes referred to as a *calligraphic writer* or *stroke writer* display. The vector refresh CRT draws in a manner similar to the vector storage display. Because of this feature, the image produced has the same characteristics as the vector storage, i.e., clear, sharp, and no jaggies. See Figure 4-8. The vector refresh CRT also continually refreshes the image similar to the raster CRT. This allows selective erasure of images, eliminating one of the disadvantages of the vector storage CRT.

Refreshing is accomplished by continually repainting the image on the screen, thus creating the major disadvantage of the display. On a complicated drawing with many lines and elements, the constant redrawing of the image causes the image to flicker on the screen. This flickering can be very annoying to the operator. On simple drawings this is not a problem, and does not become a problem as long as the redraw rate is below the flicker threshold of the human eye.

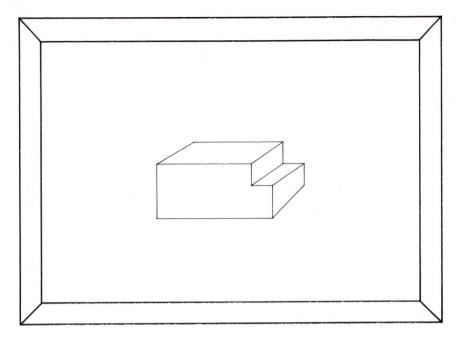

FIGURE 4-8 A vector refresh tube draws from point to point as does the storage tube. The lines must be constantly refreshed.

Color CRT Display Terminals

Some industries have a need for color in designing. The electronic industry's design of chips is much more easily performed by using colors. CAD/CAM applications and engineering functions, such as stress and strain on parts, are much more vividly shown on screen with the use of color. The amount of stress on a part can be represented by the brightness of the color red, for example. For these and other reasons, the use of color display terminals is increasing.

Most color CRTs used in CAD are raster displays. This trend will continue for a number of years. By using three different guns, one for each of the primary colors red, green, blue, it is possible to create images from a palette (A *palette* is a range of colors) of more than 1 million colors. However, the total number of colors that can be displayed at one time on the screen is much less than the total number available to the designer.

Although screen resolutions in the range of 1024 by 800 are not uncommon, dramatic improvements are taking place. The highest resolution available on a color raster graphics terminal is 2048 by 2048 pixels. Line quality is as good as a vector system and can be used as a replacement of stroke writer displays. These display devices can generate as many as 256 colors simultaneously from a palette of over 250,000 colors.

Most microcomputer-based CAD systems support a number of different display devices. Most of these display devices are raster and IBM compatible because the IBM PC, XT, and AT have become the *de facto* standard for microcomputers. For this reason, most micro-based CAD software is written for IBM computers. However, CAD software will run on most IBM compatibles such as Epson, Compaq, Zenith, and others. In addition, the software may run on IBM clones, which are IBM PC look-alike computers.

Graphic Display Cards

There are four standard graphic display cards that have been developed to support IBM microcomputer hardware: (1) Monochrome, with 720 by 348 pixels of resolution; (2) Color Graphics Adapter (CGA), with 640 by 200 pixels of resolution in black and white and support of four colors from a palette of 16 in the color mode; (3) Enhanced Color Graphics Adapter (EGA), with 640 by 350 pixels of resolution and 16 colors from a palette of thousands; (4) Professional Graphics Adapter (PGA), with 1024 by 800 pixels of resolution and 256 displayed from a palette of 16.8 million. Some monitors will only run on a certain type of graphics monitor and some graphics cards will support more than one type of standard.

Flat CRT Technology

CRTs are too bulky when compared to the size of today's powerful but small computers. Flat CRTs would be less bulky, lighter in weight, and less costly to operate. By 1990, it is expected that flat panel displays will be challenging the market of today's displays. Some flat displays are available today, but their high cost prevents widespread use. (Figure 2-4 in Chapter Two shows a flat terminal display marketed by IBM.) Flat terminal technology includes plasma display, liquid crystal display, and flat CRT technology.

FIGURE 4-9 Dual raster display terminal used to maintain visual continuity. *(Courtesy of Bausch & Lomb)*

FIGURE 4-10 Separate display device used to show alphanumeric information only. This allows more screen area for the drawing. *(Courtesy of McDonnell Douglas Automation Co.)*

Dual Display Terminals

An emerging technology related to display devices is the use of dual displays for CAD. This technique uses two CRTs to display graphics. Figure 4-9 shows a dual raster display option available on one type of CAD system. The dual display allows the operator to view an entire image on one screen, while zooming takes place on the other screen to perform detail work. Dual screen technology allows the operator to maintain visual continuity among views, details, layers, and 3-D views.

Alphanumeric CRT

Using CAD for drawing involves the frequent use of messages and alphanumeric information between the operator and the computer. The reason for this is that the software has been written in such a way that the computer will display messages during the drawing process. Messages provide the operator with improved communications between the computer and the operator. On many CAD systems, these messages are put on the screen along with the drawing, thus decreasing some of the drawing area.

To eliminate alphanumeric messages on the drawing screen, some CAD systems have a separate CRT used only for the display of alphanumerics. Any messages between the operator and the computer are then displayed on the alphanumeric terminal, leaving more room on the graphics terminal for drawing. Figure 4-10 shows a terminal with two separate screens. The screen on the left is the alphanumeric CRT.

Display Device Summary

When evaluating a CAD system, you will have to decide which type of display best suits your needs. If you need superior resolution without jaggies, the vector storage device or vector refresh display device may be the best. However, for most CAD uses and when color is a consideration, the raster display is the most popular choice. The complexity of certain designs may make a dual screen display a choice to consider. When an operator spends many hours looking at a screen, the type of display device on a CAD system is very important for preventing fatigue and for maximizing the operator's efficiency.

INPUT DEVICES

Another major hardware component of a CAD system is the input device. The *input* device is a mechanism used to interact with the computer by entering information into memory. Many types of devices are used to enter information, and they can change from one system to another. CAD systems also vary in the number of input devices used for interaction. Multiple devices provide versatility and speed of operation. A CAD system limited to one input device for interaction is not a very efficient system. For this reason, CAD systems generally have more than one input device.

Locator Devices and Menus

Before getting into the specific types of input devices, it would be helpful to become familiar with the locator device and menus used with some types of input devices. The graphics tablet is the most common type of locator device used in CAD. Graphics tablets come in a variety of sizes, two of which are shown in Figure 4-11. Most tablets have a grid of wires running beneath their surface that senses the position of the input device. This position is then displayed on the monitor in the form of a screen cursor.

Tablets

The graphics tablet can be used for three different functions. One of those functions is to locate a screen cursor; thus its name of *locator device*. Typically, a *graphics tablet* is a flat surface over which an input device is used to move the screen cursor. A *cursor* is a small flashing line, box, or cross hairs that appears on the CRT. Moving the mouse input device from right to left on the table causes a corresponding move of the cursor on the CRT. See Figure 4-12.

Graphics tablets can also be used for menu selection. The meaning of the word menu in CAD is derived from the same word that is used in a restaurant. A restaurant has a menu that displays all the choices of food and beverages that it serves. The customer looks at the menu and chooses from the selections listed. In a similar way, a drafter looks at the CAD menu

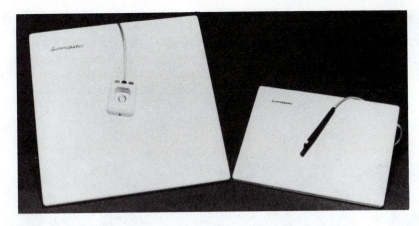

FIGURE 4-11 Two sizes of graphics tablets. The input device on the right is a stylus, and the input device on the left is a puck. *(Courtesy of Summagraphics Corp.)*

and chooses the drawing operation necessary to complete a task. Unlike some restaurants, the CAD system will serve you immediately. For most menu items the operator is served promptly after selecting the menu option. This is what makes CAD such a powerful tool when compared to traditional drafting. A *menu* then, is a list of drawing and support commands that the CAD system has available to the user. A menu may appear on the CRT, or it may be located on the graphics tablet as shown in Figure 4-13.

To use a menu on a tablet it is necessary to move the input device to a position over the selection. Through the input device, the operator then signals the computer as to the selection wanted, Figure 4-14. This menu selection actuates the computer to perform a specific function controlled by

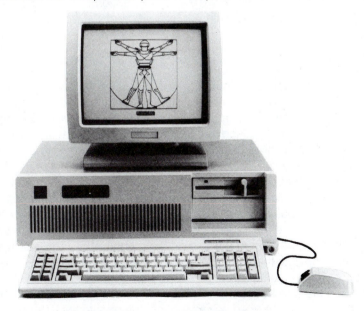

FIGURE 4-12 Mouse used to control the cursor. *(Courtesy of Robo Systems.)*

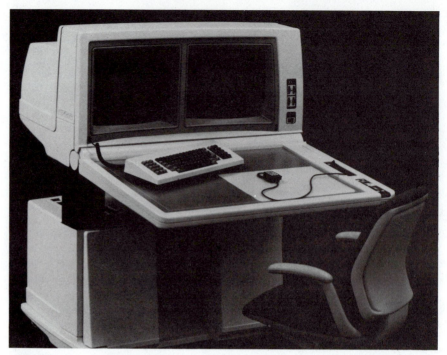

FIGURE 4-13 Menu commands can be selected from the graphics tablet located beneath the puck. *(Courtesy of Intergraph Corp.)*

the operator, such as drawing a circle of a certain diameter and locating it on the drawing. To change the drawing function from drawing circles to drawing lines, for example, the operator would move the input device to the menu location for lines and make the selection. This will now actuate the computer to draw lines to the specification of the operator.

Menus

Sometimes, menus appear on screen instead of on the graphics tablet. Screen menus can be located at the top or bottom of the CRT or to either side. The remaining space is used for the drawing. Having the menu on screen can be a disadvantage in that it takes away from the size of the screen used for drawing. However, an advantage of screen menus is that the drafter's eyes are always looking at the screen instead of constantly switching attention from the screen to a menu located off screen.

Selecting drawing options from a screen menu is accomplished by moving the input device over the graphics tablet causing the screen cursor to locate over the menu item. Some systems use a light pen to pick the menu by positioning the tip of the pen on the menu option. The operator, through the input device, then signals the computer of the menu choice. Sometimes, the list of menu items on the screen has a number or letter before the item; in which case, the keyboard can be used to choose the menu item. Screen menus can also be used by moving the cursor onto the screen choice which

FIGURE 4-14 The user positions the crosshairs of the input device over the menu item and signals the computer by depressing a button on the puck. *(Courtesy of Bausch & Lomb)*

will automatically activate that command from the menu. It is not uncommon for CAD systems to employ all three methods of screen menu selections described. Figure 4-15 shows a screen menu.

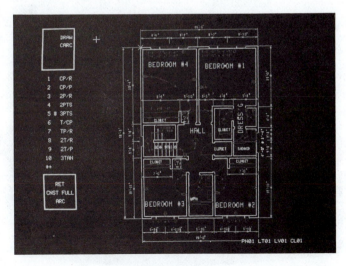

FIGURE 4-15 A screen menu is located on the left side of the screen. Menu choice #5 is activated on this menu as shown by the small light rectangle. The + located in the upper left of the screen is the cursor. *(Courtesy of Bruning CAD)*

Pull-Down Menus and Icons

Another common method used for menu selection is through the use of pull-down menus, most commonly found on Apple computers. This method of menu selection is becoming more popular even on IBM microcomputer CAD software. Figure 4-16 is an example of a pull-down menu. A pull-down menu will temporarily cover part of the drawing but will disappear after the menu item is selected. Figure 4-16 also reveals a menu of icons at the upper right section of the screen. *Icons* are graphic representations or pictures of words. For example, the command to create a circle might have a picture of a circle instead of the word "circle." Pointing to the picture or icon of the circle with the input device will activate that menu item and allow the operator to draw circles.

Digitizers

A digitizer is an input device that can also be called a graphics tablet; that is, a digitizer and a graphics tablet can be one and the same device. The Intergraph CAD system has a large work area in front of the terminal, as shown in Figure 4-17. This work surface is a digitizer board that can be used for digitizing a drawing or as a graphics tablet for menu selection.

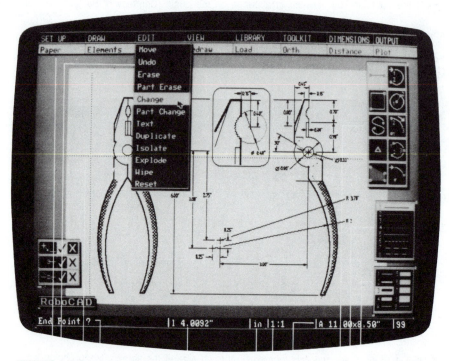

FIGURE 4-16 Screen display showing a CAD system that uses pull-down menus and icons for some drawing functions. *(Courtesy of ROBO Systems Corp.)*

FIGURE 4-17 The large work surface in front of the dual screens can be used as a digitizer board or as a tablet for menu selection. *(Courtesy of Intergraph Corp.)*

A tablet used only as a locator device does not need to be as accurate as a tablet used for digitizing. Typical accuracies of digitizers are .01'' or more, but accuracies of a tablet not used for digitizing can be as low as .05''.

Digitizing is used to input a drawing produced on paper into the graphics system. This is accomplished by taping the drawing onto the digitizing tablet and using the input device to locate end points of lines, arcs, and centers of holes. Labels and dimensions are added after the drawing has been digitized into the computer. Because the drawing has to be located on the digitizer board, digitizer boards come in many sizes to accommodate different sizes of drawing media.

Automatic Scanners

The increase in the use of CAD is creating a problem for many companies. The problem is how to convert drawings created with traditional tools to a format that can be used with a CAD system. Digitizing drawings manually can take many hours or, in the case of large format drawings, days. However, advances in document scanning and pattern recognition technology have made automatic or semiautomatic document conversion possible. These automated systems can convert engineering and technical drawings, maps, and other large hard-copy documents to a format that can be used with a CAD system in much less time than it would take using manual digitizing techniques.

Keyboard

The most common type of input device used in CAD is the keyboard. The keyboard is arranged in the same order as a typewriter keyboard, although many other special or programmed keys are added to the keyboard for drafting or computing.

The keyboard is used to interact with the computer to produce a drawing by entering alphanumeric data (letters and numbers). Keyboards are used for screen menu selection as described previously, adding labels to drawings, computations, keying special characters, and using programmed keys for some drawing functions. Because a keyboard is used so often, a knowledge of typing can be very helpful to a drafter.

Figure 4-18 shows a keyboard with control dials, joystick, calculator pad and special keys for drawing.

Light Pen

The light pen was one of the first input devices developed for CAD. The *light pen* is an input device used to make point and menu selection by sensing light emitted from the picture elements on the CRT. The light pen is moved about on the screen and point choices are made or menu items selected. One disadvantage of the light pen is that an operator can become arm weary from holding the light pen up to the screen. For this reason and others, the light pen is not being used as often as it has been in the past.

Figure 4-19 shows an operator using a light pen as an input device to locate the end point of a line.

FIGURE 4-18 Keyboard used in CAD showing various input devices. *(Courtesy of Megatek Corp.)*

FIGURE 4-19 Using a light pen as the input device for drawing. *(Courtesy of the International Business Machines Corporation)*

The Stylus and Graphics Tablet

The stylus is another common input device that is used with a graphics or data tablet. As shown in Figure 4-20, the stylus is very similar in appearance to the light pen. However, that is where the similarity ends.

Do not confuse the stylus with a light pen; they are two different input devices. As described previously, the light pen is pointed at the screen. The *stylus* is a locator device that is held like a pen and moved over the tablet to control the location of the screen cursor. It can also be used to make menu selections from the tablet. This is usually done by placing the tip of the stylus over the menu selection and digitizing it. Digitizing the menu item or point selection is done by pressing down on the tip of the stylus or by depressing a small button on the stylus. Either action activates the stylus, thus actuating the computer to perform the menu operation chosen. The system can also be used to choose screen menu items indicated by the cursor which is controlled by moving the stylus over the tablet.

The Puck and Graphics Tablet

The *puck* is a locator device used with a graphics tablet to control cursor location. Pucks come in many shapes and sizes, with varying numbers of buttons located on their surface. Some pucks also have cross hairs for accurate digitizing or for menu selection, Figure 4-14. Sometimes, a puck is referred to as a cursor by some CAD manufacturers. This is a misnomer. If you see a puck referred to as a cursor, do not confuse this reference to a locator device with the screen cursor.

FIGURE 4-20 Using a stylus to make menu selection from a tablet and for cursor control. *(Courtesy of CalComp)*

The puck functions in the same fashion as the stylus. Moving the puck across the tablet causes a corresponding change in the position of the screen cursor. The puck can also be used to select items on a menu located on the tablet. This is accomplished by moving the cross hairs of the puck over the menu item desired. The proper button on the puck is then depressed, activating the selected drawing function for the operator. The puck can also be used to select menu items on the screen by controlling the location of the screen cursor. The screen cursor is moved over the menu item on the screen, automatically activating that item, or one of the buttons on the puck must be depressed to activate the selection.

Mouse

Another type of input device that can be used is the mouse. A *mouse* is an input or locator device used to position a screen cursor. There are two types of mice in common use: mechanical and optical. The main advantage of the mechanical mouse is that no special tablet has to be used for its surface. The mechanical mouse operates on any smooth tabletop. Figure 4-12. The mouse has rollers located on the base with circuitry connected to them. Moving the mouse causes the rollers to turn, which then moves the cursor on the screen for drawing and menu selection. The mouse can be picked up and placed in any spot on the tabletop and the cursor location on the screen will not change, because the rollers in the base did not turn.

Another type of mouse used in CAD is the optical mouse, as shown in Figure 4-21. This mouse uses mirrors in the base to locate its position over a small reflected plate set on a tabletop. The mouse usually has buttons on its surface which are used to pick or activate points or functions, in much the same way as the puck is used.

FIGURE 4-21 An optical mouse input device that uses a reflective surface to control cursor movement. *(Courtesy of Summagraphics Corp.)*

Joysticks

The *joystick* is another input device used to control screen cursor location. The joystick can be moved to the right or the left, forward, backward and diagonally, causing a corresponding move of the cursor on the screen. The speed of cursor movement is determined by the amount of tilting from vertical to which the joystick is subjected. Very accurate cursor positioning can be difficult using the joystick and is one of the disadvantages of their use. Some joysticks can also be twisted clockwise and counterclockwise to control the z axis in a 3-dimensional drawing. A joystick is shown to the right of the keyboard in Figure 4-18.

Thumbwheels

Thumbwheels are also used as an input device to control cursor location. Figure 4-22 shows a set of thumbwheels located to the right of the keyboard. One thumbwheel is used to control the x axis location of the cursor, and the other is used for the y axis location. Moving both causes diagonal movement. The cursor can then be controlled by the thumbwheels to select points on the screen or for on screen menu selection.

Programmed Function Board

The programmed function board is a very common input device used in CAD. The *programmed function board* usually is a separate key-board with rows of buttons or pressure-sensitive switches used for specific

FIGURE 4-22 Thumbwheels are another method used to control a screen cursor. *(Courtesy of Bausch & Lomb)*

programmed functions related to drawing. Figure 4-23 shows a programmed function board. The function board works in the same way as the programmed keys that are part of the alphanumeric keyboard. The programmed function board is really another method of menu selection, using rows of buttons for function selection instead of a menu on screen or on a

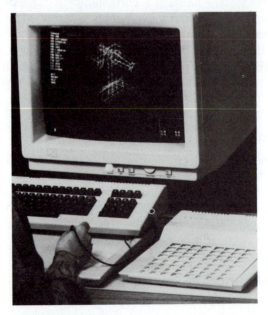

FIGURE 4-23 To the right of the operator's hand is a programmed function board used for menu selection. *(Courtesy of Applicon)*

tablet. After the button has been depressed, the system will display a menu of related commands on screen.

Some function boards have lights under their keys to signal those functions that are available to the user. Each button controls a specific function. Pressing an active button will enable the CAD system to perform a special or programmed function. If, for example, you pressed the "line" key, a menu would appear on screen listing the various methods used to draw a line. The user would then select the method to be used to draw the line from the list of menu options.

Other Input Devices

Other CAD input devices are available, but they are not in wide-spread use. A trackball is one such input device, and it is similar to a device used in arcade games and home computer games. Speech recognizers are other input devices that are not commonly used, but they may become more common in the future. These devices are trained to the operator's voice and the drafter can then make selections by voice command. Control dials are other input devices that are used on some CAD systems to control zoom and pan functions.

Whatever input device is used in a CAD system, the trained operator should be familiar with as many types as possible. Most systems have more than one input device to make the system more versatile and time efficient. The trained CAD operator should not be misled by a manufacturer's own names for input devices. For example, a board with rows of buttons used to control certain drafting operations is still a programmed function board whether the manufacturer calls it that or "Acme's super-duper button board."

OUTPUT DEVICES

After producing a drawing on screen, a more permanent method of viewing the work must be used. Traditional drafting practices use some type of paper medium to produce a drawing, so images are not easily lost. In CAD, methods have been developed to produce permanent drawings of images created on screen. These devices are considered to be *output devices,* and they are grouped according to the method by which the drawing is reproduced.

Electromechanical Pen Plotters

Electromechanical pen plotters produce a finished inked drawing of high quality, neatness and accuracy, in multiple colors, that can be created on such varied drawing media as vellum, Mylar®, and transparency. *Electromechanical pen plotters* are output devices that use pens mounted in blocks or holders that move across the drawing medium to produce a permanent image. Pen colors, type, and thickness can be automatically

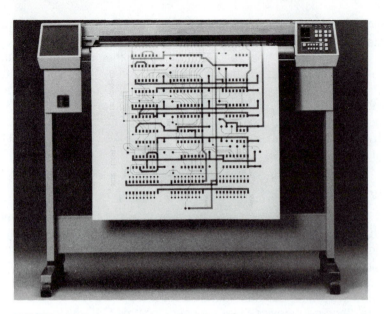

FIGURE 4-24 E size microgrip drum plotter. *(Courtesy of Hewlett-Packard)*

changed by the plotter. Pen types include ball point, felt, and wet ink. Pen plotters typically can use between one and 14 pens, and the use of the plotter paper can vary from A size to larger than E size. Two types of electro-mechanical pen plotters are drum plotters and flatbed plotters.

Drum Plotters

A *drum plotter* uses pens mounted in a block that moves across the paper along a bar. The paper is mounted on a drum that rotates, providing the second axis of motion.

The paper is held to the drum in one of three fashions. The model shown in Figure 4-24 is an example of a *microgrip* or *pinch grip* drum plotter. The paper is held by pressure of pinchwheels. A sprocket-fed drum plotter uses special paper that has holes punched in two edges. The plotter has sprockets that control the movement of the paper. This type of plotter is usually slower plotting than the microgrip plotter. The third type of drum plotter is the belt bed plotter, as shown in Figure 4-25. The paper is taped onto the belt that rotates as the pens move across the paper.

Flatbed Plotters

The *flatbed plotter* differs from the drum plotter in that the paper remains stationary while the plot is being made. The pens can move in two dimensions, unlike drum plotters which can move in only one. The paper is typically held to the flatbed through an electrostatic charge or by vacuum. Figure 4-26 shows a photograph of a flatbed plotter.

FIGURE 4-25 A large, 4-pen belt bed plotter. *(Courtesy of CalComp)*

Flatbed plotters are used for engineering drawings, templates, and artwork for electronic PC boards and chips. The pen can also be replaced by a knife or light to produce electronic circuit artwork.

Pen plotters are rated by their plotting speed and their accuracy or repeatability. Plotting speeds are normally in the range of 5 IPS (inches per second) to 30 IPS. *Repeatability* is the measure of the accuracy of the plotter to retrace an element such as a line or circle. Typical repeatability measures for pen plotters are in the range of .005 to .001 inch. Of course, the higher the plotting speed and accuracy of the plot, the more expensive the plotter. On a typical microcomputer-based CAD system, the plotter may be the most expensive single item for a workstation. However, it is not uncommon for one plotter to be shared by a number of workstations.

Plotters may also have buffers that can be used to store the plot data while the plot is being performed. Some buffers will allow multiple-plot data storage so that more than one drawing can be loaded into the plotter. This will free the computer and the operator so they can continue their work without interruption.

All pen plotters are relatively slow when compared to other output devices. They are the most popular type of output device used because of their versatility, reasonable cost, and superior quality of print. Because of its high quality, the pen plotter is used primarily for producing permanent copies of drawings produced on CAD.

Electrostatic Plotters

An *electrostatic plotter* produces a drawing using a method which is completely different from that of a pen plotter. The drawing is produced electronically by charging the paper and placing a toner onto the surface

FIGURE 4-26 A large, 4-pen flatbed plotter. *(Courtesy of CalComp)*

which then turns dark at the point of contact. The paper is charged in the form of many tiny overlapping dots that are placed from 100 to 508 dots per inch (DPI). The paper is fed from a roll under a matrix that charges the plotting medium. Plotting speeds are very fast, from 0.25 to 1.5 inches per second, but these plotters are usually more expensive than electromechanical plotters. Figure 4-27 shows an electrostatic plotter. The latest advance in Electrostatic Plotters is the use of color.

Electrostatic plotters include such hard-copy devices as dry silver, thermal copiers or dot matrix printers. These devices are typically small in size and are used to duplicate screen images quickly. Even the most detailed drawing can be copied within seconds, usually at the touch of a button. These less expensive hard-copy devices produce copies which are not of high quality but which are inexpensive and fast. Figure 4-28 shows a hard-copy device, and Figure 4-29 shows an alphanumeric dot matrix printer used only for alphanumeric output.

Laser Plotting

Laser plotting is a nonimpact electronic method of printing that uses a laser beam for image transfer using dry toner. A typical resolution is 300 dots per inch that can produce detailed multicolored prints of good resolution and high quality. Printing speeds are similar to those produced by electrostatic plotters.

FIGURE 4-27 A high-speed electrostatic plotter. *(Courtesy of CalComp)*

FIGURE 4-28 A small dot matrix hard-copy device used to print screen images quickly for check drawings. *(Courtesy of Radio Shack, a Division of Tandy Corp.)*

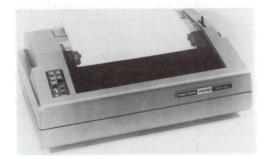

FIGURE 4-29 A 132-column, high-speed dot matrix alphanumeric printer. *(Courtesy of Radio Shack, a Division of Tandy Corp.)*

Ink Jet Plotters

The rapid growth of color graphics has been responsible for the development of multicolor *ink jet plotters*. This process forces minute ink droplets of various colors onto the drawing medium, resulting in a multicolor print. A typical ink jet plotter deposits 200 ink droplets per inch in a range of 4 to 4000 colors and shades. Plotting speeds are slow, but the final print produces a good-quality multicolor print. These Plotters are very useful for solid modeling applications because of their shading capabilities and the density of color produced.

Photo Plotters

Photo plotting is a photography-based plotting technique used primarily for printed circuitry. A *photo plotter* draws a pattern by exposing a light-sensitive medium. The size of the light beam is carefully controlled, thus producing an exposed medium that must then be developed. The developed print is then used as a mask to produce PC boards.

CAM (Computer-Aided Manufacturing)

The output for a design created with CAD does not have to be documented using traditional methods such as paper. The geometric data created with the design on the CAD system can be sent through a postprocessor directly to an NC machine tool such as a lathe or a milling machine. CAM (Computer-Aided Manufacturing) uses the data of a CAD system. This data can be converted by the postprocessor to a punched tape having the machine code necessary to drive an NC machine to manufacture a part. The data could also be sent directly from the CAD system through the postprocessor to the machine tool. This is referred to as CNC (Computer Numerical Control).

Summary

The four types of plotters differ in the quality and speed of plots, price, and the types of media that can be used to plot. CAD systems ideally should have more than one output device available to the user; one to produce fast, inexpensive check copies, and one to produce high-quality final copies.

Finally, there is another type of output device called a paperless plotter that is being developed. This device is used to speed checkplot production and reduce paper and ink costs. These devices can display a D size drawing with line widths as fine as 10 mils and in as many as 4096 colors. The image can be created in less than one minute and can be used for checkplots for mechanical drawings, integrated and printed circuit designs, and seismic data.

MEMORY DEVICES

The final piece of hardware that makes up a CAD system is the *memory device*. This peripheral device is used to store and retrieve programs and data outside of the computer. Memory storage has been described briefly in Chapter Three. This chapter concentrates on the memory devices used in CAD.

ENIAC, the first electronic computer, had to be programmed by rewiring the circuits by hand. The need for stored programs was soon realized. Methods were devised to store programs outside the computer on some recording medium, which could then be loaded into the computer's RAM memory when needed. Because a memory device is used to *output* data for storage and *input* stored data to the computer, it can be classified as an input/output device, or I/O device.

Magnetic Tape Storage

Two of the first methods developed to store programs were with punched cards and paper tape. These were largely replaced by magnetic media in the 1960s. *Magnetic tape* was one of the first types of magnetic media developed for program storage. Tape storage is still in large use today in CAD, but mostly for archiving drawings, Figure 4-30. It has been replaced, except for archiving, because drawings stored on disk can be retrieved much quicker than on magnetic tape. To find a certain piece of data on tape, one must wind through all the intervening tape. This is called *sequential access*, and it is a much slower process of retrieving data than using disk storage.

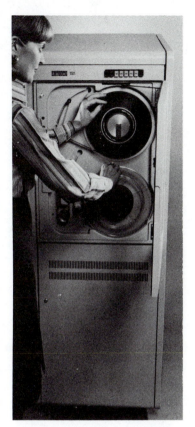

FIGURE 4-30 Magnetic tape drive. *(Courtesy of Digital Equipment Corp.)*

Disk Storage

Disk storage was developed to solve the problem of tape storage. A *disk* is a circular, flat piece of plastic or metal having the same type of magnetic coating found on tape. With a disk, the recording head can quickly search out a piece of data in the same manner that a tone arm of a phonograph record can be lifted and placed onto the desired position on a record. This is referred to as *random access*, and it is a much faster method of information retrieval than sequential access.

Floppy Disk Storage

Two common sizes of disk storage are the 8″ disk or floppy disk and the 5-¼″ diskette or minifloppy disk, Figure 4-31. The term *floppy disk* comes from the fact that the disk is flexible and not rigid as is a 45-RPM record.

Both sizes of disks have a vinyl jacket with a small cutout section, called a *head window*, to expose the disk to the disk drive head. The disk spins in the disk drive at about 300 revolutions per minute. When a command is received to retrieve data, the read/write head is lowered onto the

FIGURE 4-31 8″ and 5-¼″ floppy disks used to store data.

disk through the head window, and the magnetically stored data are searched and then read by the head, and input into the computer. The head is also used to magnetically write information onto the disk. When a command is received to store data located in random access memory, the head is lowered and the data is converted into magnetic form for writing and storing onto the disk. The function of the disk drive head is very similar to the play and record head on a common tape recorder or cassette tape player. The head itself is quite different, although the function is the same.

Figures 4-32 and 4-33 show a floppy disk and a cutaway of the vinyl jacket. The cutaway shows how data are stored on a disk. Data are grouped into concentric circular rings or tracks which can be divided into arc-shaped areas called *sectors*. The amount of data stored depends on the number of these tracks and sectors and the density of the recorded information. How dense this data can be stored determines whether the disk is single, double, or quad density. Some disk drives have two read/write heads; one for each side of the disk. This doubles the storage capabilities of the disk.

Microdisk Storage

The newest member of the disk family is the *microdisk*. The microdisk is housed in a sealed plastic package that does not have to be handled as carefully as a floppy disk. These disks have an even greater storage capacity than the 5-¼″ minifloppy; in the range of 500 K per side. This compares to about 250 K on the minifloppy, and about 1 Mb on the 8″ floppy. Another advantage of this disk is the fact that the drives run at 600 revolutions per minute; twice the speed of the minifloppy. This means that data transfer and storage is much faster.

The most common microdisk size is 3-½″. As this technology advances, microdisks will become more common and will approach the speed and storage capacity of the much more expensive Winchester disk drives. Figure 4-34 shows a microdisk.

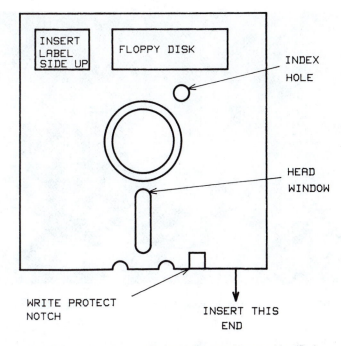

FIGURE 4-32 A floppy disk and important features identified.

Disk Densities

Disks and disk drives are grouped according to the size of the disk
(8″, 5-¼″, or microdisk), the density of the disk that can be read, and
whether the disk reads on one or both sides. For example, *a dual 8″ DSDD*

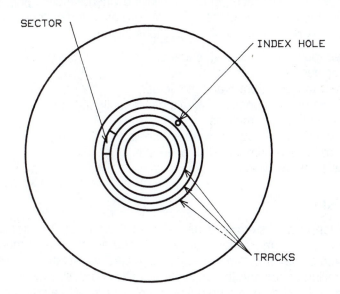

FIGURE 4-33 Cutaway showing the magnetic disk and the format used to store data.

FIGURE 4-34 3-½″ microdisks sealed in a rigid plastic cover. *(Courtesy of Bruning CAD)*

disk drive is one that has two disc drives (dual) and reads and writes on 8″ double-sided, double-density disks. The highest-density disks can store more than one megabyte or 1 million bytes or characters.

Disk Handling

Certain precautions must be followed when handling floppy disks and minifloppy disks. Most of these precautions are listed on the envelope in which the disk is stored. These precautions include:

1. Never touch exposed disk surfaces.
2. Never bend or fold the disk.
3. Protect the disk from magnetic fields.
4. Maintain proper conditions: avoid direct sunlight, high or low temperatures, and high humidities.
5. Use a soft-tipped pen to write on labels.
6. Do not force or jam into disk drive slot.
7. Make backup copies of disks.

Hard Disks

Floppy disk storage is used for most small CAD systems. Hard disks are used with more powerful systems and are becoming more common in even some of the smaller standalone CAD workstations. This is because of falling prices and the advantages of hard disk storage over floppy disk storage. *Hard disks* are rigid aluminum platters, with a magnetic coating, that spin in an airtight enclosure. A hard disk spins much faster than a floppy disk (as fast of 3600 RPM), making disk storage and retrieval much faster. The read/write heads do not touch the surface of the disk but ride very close to the surface. Storage capacities are also much greater than for floppy disks. The Winchester hard disk drive is one of the most common types of disk drive used in CAD.

Some hard disk drives are competitive with floppy disk drives in price, and come in removable cartridges. Others are sealed in airtight enclosures and cannot be removed, so that the data must be backed up on another hard disk or on floppies or tape. Data backup is necessary in the event of a disk crash or system failure. A *crash* is a mechanical failure of the disk drive that causes the information stored on the disk to become unreadable by the disk heads, thus resulting in the loss of that data. Backup is usually accomplished by using a floppy disk.

Disk Packs

Very large CAD systems, operating from minicomputers or mainframe computers, use different storage media than are used by smaller systems. These systems generally use disk packs and magnetic tape as a backup. A *disk pack* is a number of rigid, magnetic disks stacked together as a unit or pack. Because they are not easily removed, they must be backed up with another medium, such as magnetic tape (described in detail earlier in this chapter). Figure 4-35 shows a disk pack.

Bubble Memory and Others

Bubble memory is a recent development that is faster than disk storage. This technology uses a chip to store data, but information will not be lost when the power is turned off as it is with random access memory storage. Increased use of bubble memory is expected as the technology

FIGURE 4-35 A 256-Mb hard disk drive, and removable disk pack. *(Courtesy of Digital Equipment Corp.)*

develops and improves. Bubble memory technology will improve and become cheaper in cost, making it a more common method of storing data in the future. With this developing technology, storage media will continue to become smaller in size.

The *microdisk* is the newest step in the shrinking size of the floppy disk for storage. Another technology being examined for storage of data is *optical recording*. Lasers are used to write and read information on disks similar to the home video disk systems for movies. Another technology being developed is the *3-dimensional memory chip*. This memory chip has memory circuits stacked one on top of the other, allowing for much denser memory storage per chip.

Bernoulli boxes have also been developed for data storage and retrieval. This device provides fast access speed and large storage capacities like a hard disk. It is removable like a floppy disk for use as data backup and storage . Finally, one of the more recent developments has been the RAM disk or hard card. This technology utilizes RAM memory chips on a special board for rapid storage and retrieval of data. Large programs can be run from the RAM disk to decrease access and retrieval times. Of course, this memory is volatile and new data must be saved on floppy or hard disk before the computer is turned off.

SOFTWARE

In order for a computer to become a CAD system, it is necessary to use a program that performs drafting tasks. The drafter/designer is not trained in programming, and the amount of time needed to program a computer to perform drafting is expensive. For these reasons, when assessing the components necessary to make a computer a CAD system, one must include the software. The software needs are of two types: operational and application.

Operational Software

Operational software, or system software, controls the operation of the computer (CPU) and such peripheral devices as the plotter, disk drive, and CRT. This software is not visible to the CAD operator. It controls the general operation of the CAD system, but is invisible to the user.

Application Software

The *application software* is what converts the computer into a CAD system. It is the program written to produce drawings by computer in a format understandable to the trained drafter. This software is developed by the manufacturer or vendor, and it is marketed as a package to the public. Application software can be developed for drafting in general. However, most application software is created for two or three types of applications, such as mechanical, architectural, electrical, and so forth. It is not uncommon to find software developed specifically for one type of application, such as printed circuits. This does not mean that other types of drawings could not be created with this software; it means only that many drawing tasks are developed in the software that are useful only for the specific application. Chapter Three lists many of the application software packages available on CAD systems.

Unlike the operational software, the application software is very important to the operator. A good application software package is easy to learn and use. It allows the operator to be creative and to forget that a computer is being used for drawing. This type of software is referred to as being user friendly. A *user-friendly* application software package is one that is easily learned, causes the least frustration, and increases the operator's efficiency. As an operator, one of the most important features that a CAD system can have is a user-friendly software package.

User Software

User software is software created by the user to enhance or upgrade the application software. Basically, two methods are used for accomplishing this. In one method, the operator makes a drawing of a part, such as a transistor, that will be used repeatedly on a drawing. This drawing of the transistor will then be stored as a symbol or a template. Whenever a transistor is needed on a drawing, retrieving this symbol by pressing a few buttons will result in a tremendous time savings. Such symbols can be created by the user for any purpose. This is a feature that virtually all CAD systems offer.

The second method used for an operator to create software is through parametric programming or macros. A *parametric program* is a group of English or programming language-like statements chained together to perform a function. These programs call on drawing functions from the application software to create part of a drawing. Many of the more sophisticated CAD systems provide parametric programming for user-created software.

THE USER

The final part that makes up the CAD system is the most important: the user. *The user* is the most important because, without this person's proper knowledge and training, the use of a computer for drafting would not offer many advantages over manual drafting.

Operators trained in the fundamentals of drafting are necessary to make the computer perform at its optimum capacity. Ideal CAD operators are drafting technicians with CAD training or experience. These operators are more than just users of a software package. They provide the background, creativity, and knowledge necessary to be more than just users. A trained CAD operator is a drafting-design technician who makes a CAD system achieve its potential: a tool to increase productivity and relieve the drafter-designer of repetitious and costly tasks.

Chapter Four GLOSSARY

Alphanumeric—consisting of numbers, letters, and special characters which are input to the computer by means of the keyboard.

Application software—the programmed instructions that allow the user to employ a computer for a specific task or tasks in a format understandable to the operator.

CAD—computer-aided drafting or computer-aided design.

CAD operator—a drafting technician with CAD training or experience.

Crash—mechanical failure in a disk drive or other piece of computer equipment or software program, usually resulting in some loss of stored data.

Cursor—a small flashing line, box or cross hairs, used to locate positions on a CRT.

Disk—a circular flat piece of aluminum or plastic material coated with a thin layer of magnetic material used to store data. Sometimes called "floppy disk," "minidisk" or "diskette," "microdisk," "hard disk," "Winchester disk," "disk pack."

Display device—a device used to display a graphic image. The CRT is the most common display device used for graphics.

Drum plotter—a pen plotter, electromechanical output device that uses pens mounted in a block that moves across the paper and along a bar with the paper mounted in a rotating drum.

Dual display—the use of two display devices to show graphics and to allow the operator to maintain visual continuity among views, details, layers, and so forth.

Electrostatic plotter—an output plotting device that produces an image by electronically charging the paper and then putting a toner over the charge to produce an image.

Flatbed plotter—an electromechanical plotting output device where the paper remains stationary on a flatbed while the pens move in two directions creating the plot.

Graphics tablet—an input device used with a puck or a stylus to locate a screen cursor, to digitize a drawing or to select menu items.

Icon—a picture or graphic representation of a word.

Ink jet plotter—an output plotting device that forces minute droplets of ink onto the paper to produce an image.

Input device—a mechanism used to interact with a computer or to input data.

Jaggies—phenomena, associated with raster display terminals, which cause angled lines to appear jagged or stair-stepped. Also known as *aliasing* or *stair-stepping*.

Joystick—an input or locator device used to control the position of a screen cursor.

Laser plotter—an output plotting device that uses a laser beam and toner to produce an image on paper.

Light pen—an input or locator device used to make point and menu selections by sensing light emitted from the picture elements on a CRT.

Locator device—apparatus used to locate or position a screen cursor.

Magnetic tape—a thin strip of plastic film coated with a thin layer of magnetic material used to store data and to archive drawings.

Memory device—a peripheral device used to store programs and data outside of the computer. Examples include magnetic tape, floppy disk, hard disk, and so forth.

Menu—a table or list of drawing and support commands from which a CAD operator can choose to produce a drawing.

Mouse—an input or locator device used to position a screen cursor.

Operational software—software development to control the operation of the computer (CPU) and the peripheral devices. Sometimes called *system software*.

Output device—a mechanism used to produce hard copies of images or data created with a computer.

Palette—a range of colors.

Parametric program (macros)—English or programming language-like statements that call on drawing functions from the application software which are then chained together to perform a task.

Pen plotter—an electromechanical output device that uses pens mounted in blocks or holders that move across the drawing medium to produce a two-dimensional graphic image.

Peripheral—equipment or device, outside of the computer, which is controlled by the CPU.

Photo plotter—an output device that uses a light beam to draw a pattern onto a light-sensitive medium that is developed to produce an extremely accurate drawing or overlay used to make printed circuits.

Pixel—one dot or picture element that makes up part of the raster display terminal.

Programmed function board—an input device with rows of buttons that when depressed will activate a specific drawing function, such as a line.

Puck—an input or locator device used with a graphics tablet to position a screen cursor for making menu choices and for digitizing drawings.

Raster display—a display device that uses thousands of small picture elements that are refreshed thirty to sixty times a second to form an image.

Refresh—redrawing of a graphic image on a CRT.

Resolution—the clarity or sharpness of a display terminal.

Stylus—an input or locator device used with a graphics tablet to control cursor location and to select menu items.

Thumbwheels—an input or locator device used to control the position of a screen cursor, consisting of one wheel which moves the cursor in the x axis and one which moves the cursor in the y axis.

User friendly—a term used to describe hardware or software that is easily learned and allows an operator to become proficient in a short period of time.

User software—programmed instructions created by the user to enhance or upgrade the application software. Examples include user-created symbols and parametric programs.

Vector refresh display—a display device that combines the features of the raster display and the vector storage display, and one which continually refreshes its image and does not suffer from the jaggies.

Vector storage display—a display device that maintains its image for an extended period of time and does not suffer from the jaggies.

Winchester disk drive—a term associated with a hard disk made of a rigid aluminum platter with a magnetic coating, which spins in an airtight enclosure used to store data.

Chapter Four REVIEW

1. List the five subdivisions under CAD hardware.
2. By the characteristics of each display device described in this chapter, determine what type of display device is used in the CAD system you are working on.
3. List some uses for a keyboard on a CAD system other than for inputting alphanumeric data.
4. What is the advantage of having more than one input or output device available to the user of a CAD system?
5. List the three types of CRTs in use today.
6. Define the term *resolution*.
7. For what is a locator device used in CAD?
8. Describe the screen cursor that appears on the CRT of your CAD system.
9. List the input device(s) used by your CAD system.
10. List the output device(s) used by your CAD system.
11. What is the primary use of a photo plotter?
12. List the memory device(s) used by your CAD system.
13. Describe a friendly CAD system.
14. Give two examples of user-created software found in CAD.
15. Define a CAD operator.
16. Describe the location of the menu on your CAD system, and the input device used to pick menu items.

In Chapter Four, you were introduced to the various components of a CAD system. In that chapter you learned that a CAD system could be configured in many different ways. Input devices come in many forms and the majority of systems use more than one such device. This chapter will help you to group CAD systems according to the CPU, and other characteristics. The final subjects covered in this chapter describe the sequence of steps necessary to draw lines using different types of input devices. Coordinate geometry is also reviewed, since it is a necessary prerequisite before using many CAD systems.

Up to this point, the method used for actually drawing using a CAD system has only been discussed briefly. After four chapters of background information, this chapter introduces you to the sequence of steps to follow to perform a specific drawing function. This introduction is covered in a generic or general way, because of the wide variety of CAD systems being used in industry. Only three methods of program execution are discussed, because the overall majority of CAD systems fall within these following three categories: menu and input device, function board and input device, and keyboard. This chapter completes the background information associated with CAD. All of the succeeding chapters introduce the reader to the methods of program execution for various drawing functions.

OBJECTIVES

After completing this chapter, you will be able to

- group CAD systems by the type of CPU.
- explain the difference between a standalone system and a multistation CAD system.
- group CAD systems according to the method of program execution.
- list some important features to consider when evaluating a CAD system.

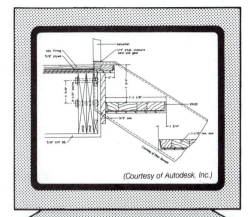

(Courtesy of Autodesk, Inc.)

Chapter Five
The Basics of a
CAD System

- describe a typical CAD system on the market by the hardware and the peripheral devices that make up that system.
- identify points using the Cartesian and polar coordinate systems.
- describe the steps necessary to draw a line on the CAD system that most resembles the system you are training on.

THE TYPES OF CAD SYSTEMS

Chapter Three describes the different classes of computers, grouped according to the CPU and computing speeds. Similar classifying can be used to group CAD systems. In Chapter Four, you learned that computers are grouped as mainframe, minis and micros. In grouping CAD systems, mainframe and mini-based CAD systems are grouped together. This is done because of the many similarities between a mainframe CAD system and a mini CAD system. Grouping them together leads to less confusion than trying to separate two very similar types of CPUs. The other group used to classify CAD systems is the microcomputer-based CAD system.

Mainframe and Minicomputer-based CAD Systems

Mainframe and mini-based CAD systems are large computers housed in an environmentally controlled room. The terminals for drawing are usually located away from the CPU in an adjoining room or at some greater distance. With this type of computer, it is possible for the terminals to be located in another city and use a modem and the telephone lines to communicate with the CPU. A *modem* is a device used to send data on telephone lines. A modem MOdulates digital information to audio sound and then DEModulates audio signals to digital form. Modems come in two types: the direct-connect modem, as shown in Figure 5-1, and the acoustic modem.

Typically, this class of CAD system has a number of terminals tied into the CPU and is referred to as a multiuser system. *Multiuser* means that more than one operator can be drawing at the same time, and that all operators can have access to any drawing produced on the system. One advantage to having access to another designer's data base (drawing) is that operators can share information which can be valuable when designing a product. A *data base* can be defined as the handling and storing of information of an entire organization, such as that produced by the engineering department, so that it can be universally used by that organization. Another multiuser advantage is that time is saved by eliminating repetitious drawings. When the drafter begins the drawing, the data base can be searched for any similar drawings that may have been produced previously by another drafter. If a drawing is found which is similar to the one to be produced, it can be called up on the drafter's terminal and modified as necessary, thus eliminating the many hours of work it would take to produce the drawing from scratch.

FIGURE 5-1 A direct-connect modem used to send and receive data over telephone lines. *(Courtesy of Radio Shack, a Division of Tandy Corp.)*

This classification of CAD systems is typically the most powerful and the most expensive, ranging in price from $100,000 and up, depending on the hardware and the number of workstations. For example, some systems can support up to 100 terminals from a central mainframe computer. Figure 5-2 shows 12 CAD workstations and peripherals attached to one mainframe CPU. The limit to the number of workstations that can be put onto a system depends upon the CPU and the amount of internal memory.

A typical mainframe and minicomputer-based CAD system runs a software package that was developed and produced by a company independent of the hardware manufacturer. This is referred to as *third-party software*. For example, Figure 5-3 shows a workstation running Intergraph Corporation software using the Digital Equipment Corporation's VAX 730 CPU.

Microcomputer-based CAD Systems

Microcomputer-based CAD systems typically are standalone turnkey systems based on a microcomputer for the CPU. *Standalone* means that each terminal is independent of another, unlike the mainframe or minicomputer CAD systems which are linked together. The terminals are independent, because each workstation has its own independent CPU, Figure 5-4. *Turnkey* means that the purchaser buys a complete package that can be turned on and run immediately after installation.

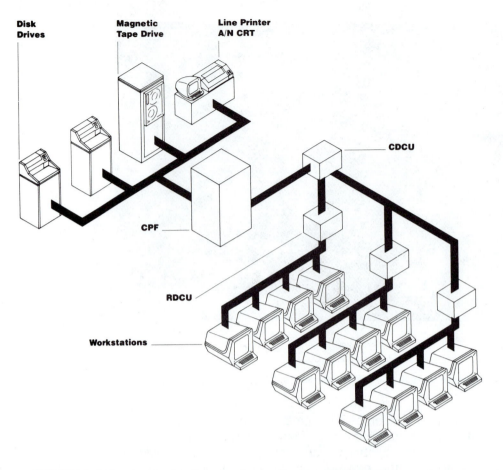

FIGURE 5-2 CAD system having 12 workstations and peripherals attached to one mainframe CPU. *(Courtesy of Vector General, Inc.)*

FIGURE 5-3 Intergraph third-party software and CAD workstation that runs on the Digital Equipment Corporation's VAX 730 CPU. *(Courtesy of Intergraph Corp.)*

A turnkey CAD system typically includes hardware, software, vendor service, and an agreement to see to it that the system works properly. The hardware includes the workstation, the CPU, and connecting cables. The software is application software needed to operate the system. The service can include installation, training, maintenance, a toll-free hot line for problems, and solutions for software bugs. The advantage of a micro-based CAD system is the low initial cost for a small number of workstations. Usually, there are no environmental restrictions with this type of system.

Another difference between a mini-based and a micro-based CAD system is the speed of drawing. Because the mini system uses a 32-bit microprocessor and has more available memory, the mini system usually draws faster. The micro-based CAD system usually uses hard disk or floppy disk drives for data storage. Mini-based systems use hard disk packs with floppy disk or magnetic tape for backup. Because of limited memory and

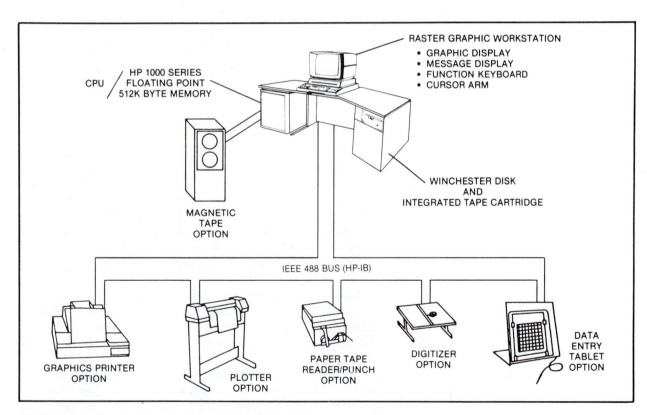

FIGURE 5-4 Typical standalone CAD system showing peripherals and workstation controlled by a microcomputer housed on the left end of the workstation. *(Courtesy of Gerber Systems Technology, Inc.)*

slower computing speeds, many design analysis functions that are run on mini-based systems cannot run efficiently on the micros. However, the trend is toward more powerful and faster micro-based CAD systems, because of improved electronics and hardware, and improved software.

Micro-based CAD systems are very common in smaller companies and in many educational institutions. Some micro-based CAD vendors are offering special packages and software specifically for the educational market. As computer technology improves, cost will continue to drop making micros more affordable to both industry and education. Appendix B shows some of the most popular micro-based CAD systems.

METHODS OF PROGRAM EXECUTION

When explaining the steps to follow to draw a line, it is impossible to describe the sequence of actions used by each CAD system on the market. However, most CAD systems and their operation can be grouped according to how the system is alerted to a specific drawing task. Most systems can be grouped under task selection as using a menu, a keyboard, or a programmed function board to select a specific drawing action. All three methods control screen cursor location through one or more cursor control devices. These input devices, described in Chapter Four, can include the light pen, stylus and tablet, puck and tablet, mouse, joystick, thumbwheels, and keyboard control.

The Coordinate System of Input

Regardless of the method used to interact with the CAD system, an operator should have a grasp of simple math and coordinate geometry. Although the computer can perform math functions much faster than an operator can, the operator must know basic math to input the information, in the correct format, for the computer to perform the calculations. To add, use a plus (+); to subtract, use a minus (−); to multiply, use an asterisk (*); to divide, use a slash (/). For example, to divide a 5.45″ line into three equal parts, the operator would key-in 5.45/3 and the computer would perform the calculation.

Coordinate geometry is important to the operator, because many CAD systems use this method of identifying points in space to create a drawing. *Coordinate geometry* theorizes that for every point in space a pair of real numbers can be assigned, and for each pair of real numbers there is a unique point in space.

Figure 5-5 shows a system of rectangular or Cartesian coordinates. A plane is divided into four quadrants by two perpendicular lines. The horizontal line is called the *X axis;* the vertical line is referred to as the *Y axis;* and the point of intersection is the *origin.* By convention, values to the left of the origin are considered to be negative, and those to the right are considered to be positive. Values upward from the origin are positive, and those downward are negative. Using these conventions, it is easy to assign any point in space a unique pair of ordered numbers. The first number would represent the X distance and the second number of the pair the Y distance. Point A in the figure would be assigned the ordered pair (2,2); point B, (−1,3); point C, (−4,−3); and point D, (1,−2). The real numbers given each point in this example are called the *coordinates*. The assigning of coordinates is simply a systematic way of locating corner points of details and objects with a CAD system.

The location of the origin on the CRT is usually in the center or bottom left corner of the CRT. Regardless of the initial location, the origin can usually be moved by the operator to a position on the screen most suitable for the drawing. Figure 5-6 shows a 2 x 4 rectangle drawn on a CAD system with the origin located in the bottom left corner of the screen. The numbered pairs represent the coordinates of the corners, measured from the origin, which were necessary to draw the object on screen. Some CAD systems will reference to the last point input and make that point the new origin. This is referred to as *Incremental* or *Relative Coordinate* entry. If your system functions in this manner, the 2 x 4 rectangle would be drawn by inputting (2,2) for A, (4,∅) for B, (∅,2) for C, and (−4,∅) for D instead of the values shown in the figure.

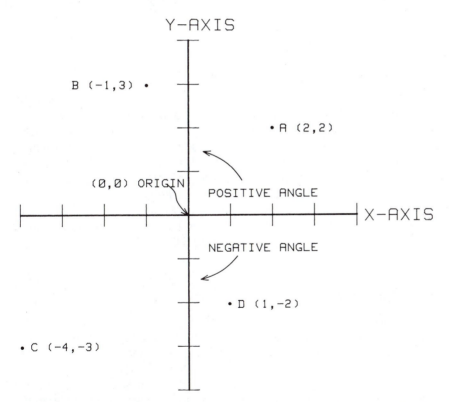

FIGURE 5-5 Cartesian coordinates, showing the important parts and the numbered pairs assigned to four different points.

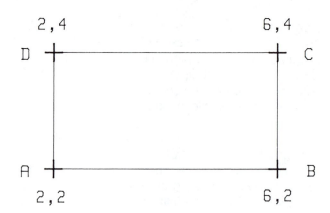

ORIGIN

FIGURE 5-6 A 2 x 4 rectangle showing the numbered pairs assigned to each corner.

In coordinate geometry, angular values are given a positive value if rotation from the positive X axis is counterclockwise, and negative if clockwise. For example, a line drawn at 45 degrees below the X axis could be measured as −45 degrees or 315 degrees (360 − 45 = 315), **Figure 5-7.**

The coordinate system just described is used for 2-dimensional drawings. For CAD systems having 3-dimensional drawing capabilities, a third axis, called a *Z axis,* must be added. The third axis would be perpendicular to the first two, intersecting at the origin. This third axis would assign a negative value to points behind the CRT screen, and positive values to points in front of the CRT screen. Drawing in 3-D would require the assigning of an ordered set of triplet numbers, with the third number being assigned the Z axis value such as (3,2,2). See Figure 5-8.

Polar Coordinates

Values can be assigned to points in space by another method called polar coordinates. The *polar coordinate* method uses the concept that any point in space can be identified by stating the angle and the distance to be traversed from the origin. A point would be identified by a pair of **numbers showing the angle first and the distance second.** Figure 5-9 shows point G as being identified as (30,4). Point G is 30 degrees from the positive X axis and a distance of 4 units from the origin measured along the

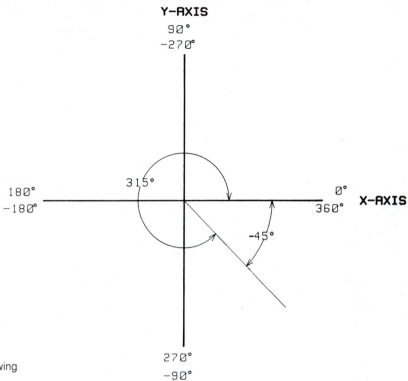

FIGURE 5-7 Cartesian coordinates showing how angles are measured.

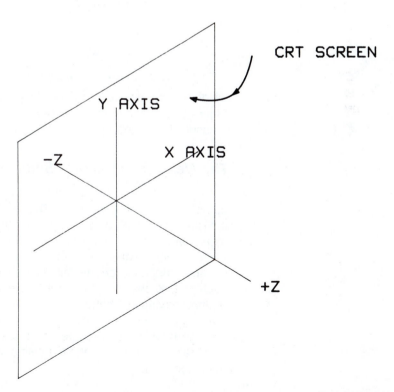

FIGURE 5-8 CRT screen showing how the Z value is assigned to coordinate points when drawing in 3-D.

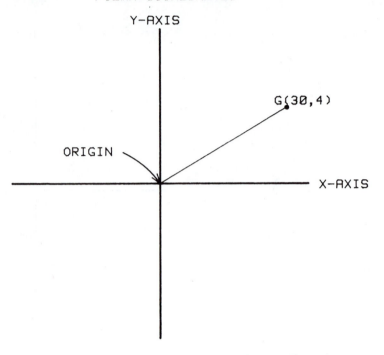

FIGURE 5-9 Details can be placed with CAD using polar coordinates which identify points by inputting the angle and the distance from the origin or the last point input.

30-degree angle. To draw the 2 x 4 rectangle using polar coordinates in Figure 5-10, point A would be (45,2), B (0,4), C (90,2), and D (180,−4) when measured from the last input point.

The Menu for Function Selection

Using a menu for control of drawing operations is one of the most common methods used in CAD. The menu may be located on the screen, on the graphics tablet, or on the programmed function board and screen. The menu, when it is located on the screen, is usually along one edge of the screen, such as the right, left, top, or bottom. Or, it may be located on a separate alphanumeric terminal as shown in Figure 5-11. Pull-down menus such as that shown in Figure 4-16 are also common. One of the disadvantages of having the menu appear on the screen is that it takes up some of the drawing area on the screen. However, an on screen menu also has an advantage over a menu located on a tablet in that the drafter's attention is centered on the screen and not split between the screen and the off screen menu location.

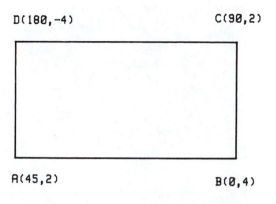

 ORIGIN

FIGURE 5-10 A 2 x 4 rectangle showing the numbered pairs assigned to each corner using polar coordinates.

On screen menu selection is accomplished through such cursor control devices as a light pen, puck, mouse, thumbwheels or keyboard. For example, the light pen is pressed against the screen at the location of the menu choice, activating that drawing function. Or the puck, mouse, or thumbwheel is moved locating the cursor to a position over the menu item on screen, which activates that drawing function. The menu on screen may

FIGURE 5-11 CAD operator selecting a menu item located on the alphanumeric terminal by using the programmed function keyboard. *(Courtesy of McDonnell Douglas Automation Co.)*

FIGURE 5-12 CAD operator selecting a menu item located on a graphics tablet. *(Courtesy of Calma Company)*

also have numbers assigned to each menu item or keys assigned to the menu option on the programmed function board. The operator can then use the keyboard or the programmed function board to select the menu item.

Menus can also be located on a digitizing tablet such as that shown in Figures 5-12 and 5-13. When the menu is located on a tablet, an input device, such as a puck with cross hairs or a stylus, is used to digitize the menu function and activate that drawing command. Screen cursor control can be accomplished through the same digitizing and input device used for menu selection, or through additional input devices such as thumbwheels or a joystick. The menu choice is selected by moving the cross hairs of the puck or the tip of the stylus over the menu item and depressing a button on the puck or stylus, thus activating that particular drawing function.

The Keyboard for Drawing Selection

Keyboards can also be used for controlling drawing functions. Keyboard selection is used on low-cost CAD systems and those marketed for educational purposes. Sometimes, the keyboard is used to supplement another method of selection, such as moving the cursor over the menu item. For example, the keyboard can be used if the menu item on screen has a number or letter associated with the selection. Pressing the letter or number chooses the menu item as does moving the cursor over the menu choice with the input device.

FIGURE 5-13 Menu located on a tablet. *(Courtesy of Autodesk, Inc.)*

However, it is not uncommon in some of the low-cost CAD systems to have keyboard input as the only method of selecting a drawing function. This is accomplished by assigning a key to a specific drawing function. For example, the letter A might be assigned the arc drawing function. Objects are described using coordinate geometry by assigning X, and Y coordinates to each point and line.

The Programmed Function Board as a Function Selector

The use of a programmed function board for drawing-mode selection is usually used in the more expensive CAD systems. The function board has a drawing task assigned to each key, and depressing a key will activate that drawing function. A key usually has a light under it to alert the operator as to whether it can be used in the current drawing mode. For example, one button might be assigned the task of drawing circles. Depressing that button and moving the cursor to the circle center or inputting the X-Y coordinates will cause the computer to draw a circle of a diameter specified by the operator. Typical cursor control devices used with a function board are a joystick, a light pen, a stylus and tablet, or a puck and tablet.

DRAWING A LINE WITH CAD

The steps to follow to draw a line on a CAD system are discussed next. An example is given for each of the aforementioned three methods used to select a drawing function. These steps are described in the most general terms, because of the great many systems on the market. These general steps should help to give you an understanding of the steps you will have to follow to draw on your particular system. Studying the manual that comes with your CAD system and carefully watching your instructor's demonstration in addition to reading these steps should provide a solid background to start drawing with CAD.

Because there are so many different cursor control devices that can be used, the steps are described without specifically naming the device. When reading through the steps, substitute the input device used on your CAD system whenever the input device is referred to. The CAD system is also assumed to be on and ready to draw. The actual start-up sequence is described in detail in Chapter Six. The reason for showing the steps used to draw a line is to familiarize you with how different CAD systems operate.

Drawing a Line Using the Menu

The steps described here use a menu located on a tablet. However, the steps will not radically differ from those used for on screen menu

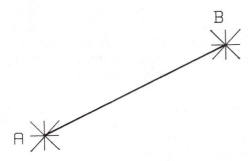

FIGURE 5-14 Drawing line AB by digitizing points A and B.

selection. The digitizing device can be either a puck with cross hairs or a stylus. Refer to Figure 5-14 while studying the following steps.

1. Move the digitizing device over the command labeled "line," "beginning of line," "separate line," or similar command, and press the button on the input device to activate.
2. Move the cursor control device to position the screen cursor in the location desired for one end of the line. Digitize the point by pressing the button on the input device. A point (A) will appear on the screen. A marker will usually appear on screen as shown at point A (*) or as a +, #, △ or some other symbol. Usually these are temporary markers that can be removed using the redraw or refresh command.
3. Move the digitizing device over the command labeled "line" if "point" had to be used to locate the first position to activate the line drawing command.
4. Move the cursor control device to the position desired (B) for the end point of the line and press the button. A line will be drawn across the screen between the two end points you located. (A,B) Many systems will have a "Rubber-band" line that will stretch between the first digitized point and the new location of the cursor on screen.

Drawing a Line Using a Function Board

Using a function board to draw a line has some similarities to using a "pick" device to choose commands from a menu. The menu used is located on the function board and is selected by depressing the button labeled for the specific drawing function desired. Depressing a button will display additional menu selections on screen. The operator then picks a specific drawing command through the input device.

1. Press the button labeled "point" on the programmed function board. This will activate the point command and display a menu on screen. The menu will give the operator a number of options as to how the first point can be activated. For example, X-Y coordinates can be keyed-in, or a point could be located at a grid point that may have been displayed on screen through a previous drawing function.

2. Move the screen cursor to the first end point with the input device. Press the button on the input device. A point will be located on the screen in the position that the cursor was located.
3. Press the button labeled "line" on the programmed function board. This will activate the line command and display a menu on screen. This menu will give the operator a number of options as to how the end point of the line can be located.
4. Key-in X-Y coordinates or move the screen cursor to the desired location for the end of the line. Press the button on the input device. A line will be drawn between the first point and the location of the cursor when the button was pressed or the location of the keyed-in coordinates.

Drawing a Line Using the Keyboard

In this example, the method of choosing a drawing function will be through the keyboard. Each key on the keyboard has a specific drawing function associated with it. Pressing a key will activate that function and allow the operator to perform a specific task through a cursor control device or through keying-in X-Y coordinates. This example will use an input device for cursor control. Input devices that can be used with this type of CAD system include a stylus and tablet, a puck, a light pen, a mouse, thumbwheels, or keyboard control. Keyboard cursor control is accomplished through the assigning of certain keys for cursor movement. Cursor keys would be marked with arrows showing the direction of cursor movement when the key was pressed. In this example, key "P" will activate the point command and key "L" will activate the line drawing command.

1. The letter "P" is pressed on the keyboard activating the point command.
2. The cursor control device is used to position the cursor at the first desired end point. The button is depressed on the input device or coordinates are keyed-in. A point (A) is located on the screen.
3. The letter "L" is pressed on the keyboard activating the line drawing command.
4. The cursor control device is then used to position the cursor at the desired point (B) for the end of the line. The button is then pressed on the input device or coordinate values are keyed in. A line (A,B) will then be drawn between the two points located with the cursor control device or the coordinate points.

HOW TO USE THIS TEXT WITH YOUR CAD SYSTEM

You might ask how it is possible to learn to draw on your CAD system by reading steps in a book that are probably different from the steps used on your system. The key point to remember is that no matter what type of CAD system is being used, there are similarities among all CAD systems.

One of these similar characteristics is the need to select drawing functions to complete a task. Virtually all systems rely on a menu or list of commands to perform a task. For example, a CAD system which uses a programmed function board for command selection is actually using the programmed keyboard as a menu-picking device. Light pens and on screen menus are simply another method of selecting drawing functions, except that the menu appears on screen instead of on a tablet. What this means is that you should carefully read through the steps listed here to perform a specific task, and then determine the steps that would be needed to perform that same task on your system. As each new menu item is covered in this text, it is very important that you find a similar menu item in your manual to determine the procedures used by your CAD system.

Another hint in using CAD for drawing is the necessity of thinking through the steps that are required for completing a drawing task before starting on the drawing. This will help you to become less confused and frustrated when solving a problem for the first time using CAD. One other important consideration is the fact that different CAD systems sometimes use different words to describe the same task. For example, if you need to erase part of a line the command might be "shield," "clip," "delete between two points" or some other expression, depending upon the CAD system being used. In other words, there is no standard word or phrase used among CAD systems to perform similar tasks such as erasing. This textbook attempts to give a number of different words or phrases that are most commonly used for a specific menu item to help you determine the comparable menu item used on your system.

When referring to the figures used with the drawing exercises in Chapters Six through Nine, drawing entities, such as lines, circles, arcs, text, dimensions, and so forth, may include a symbol resembling an asterisk, as shown in Figure 5-15. This symbol is used to show where it is necessary to digitize points, using the input device. Items are digitized through the input device as a means of communicating to the computer the location of points or entities. For example, to erase the line in Figure 5-15 you would

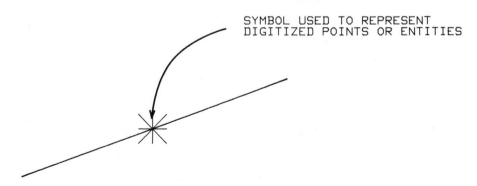

FIGURE 5-15 Digitized points are represented with a symbol resembling an asterisk.

select the proper menu item for erasing lines, then select or locate the line to be erased by digitizing any point along the line. The act of digitizing is represented in this text by the symbol located on the line in the figure.

The sample drawing exercises are described using a menu located on a tablet. As you read the examples, substitute your menu device (function board, keyboard, or screen menu) and cursor control device for the one described. Doing this will help you to understand the solution and make it easier for you to perform these tasks on your particular system.

Chapter Five GLOSSARY

Coordinate—a real number used to represent a point in space.

Coordinate geometry—for every point in space a pair of real numbers can be assigned, and for each pair of real numbers there is a unique point in space.

Data base—the handling and storing of information of an entire organization so that it can be universally used by that organization.

Modem—a device used to send data on telephone lines by MOdulating digital information to audio sound and then DEModulating the audio signal to digital form.

Multiuser—a CAD system with multiple workstations tied into one CPU with the data base of the design projects available to all users.

Origin—the point of intersection of the X and Y axes on Cartesian coordinates. Usually located at the center or bottom left of the CRT on a CAD system. An origin can usually be moved to some other position on the CRT.

Polar coordinates—a concept maintaining that any point in space can be identified by stating the angle and the distance to be traversed along the angle.

Relative Coordinates—reference is made from the last point input.

Standalone—a CAD system with each workstation independent of another.

Third-party software—software developed by a company independent of the hardware manufacturer.

Turnkey—a word used to describe a CAD system that is installed as a complete package by the manufacturer, and is ready for operation by the purchaser.

X axis—a horizontal line that passes through the origin of the Cartesian coordinates. Values to the left of the origin are negative; values to the right are positive.

Y axis—a vertical line that passes through the origin of the Cartesian coordinates. Values upward from the origin are positive; values downward are negative.

Z axis—a line perpendicular to the X and Y axes intersecting at the origin. Values behind the origin are negative; values in front are positive.

Chapter Five REVIEW

1. CAD systems are grouped according to the CPU used. Name these groups.
2. For what is a modem used in CAD?
3. Describe a standalone turnkey CAD system.
4. What are the primary differences between a micro-based CAD system and a minicomputer-based system?
5. List the three most common methods of executing drawing commands with CAD.
6. Sketch Cartesian coordinates and plot the points whose coordinates are A ($-3,-3$), B ($3,-3$), C ($6,-1$), D ($6,5$), E ($\emptyset,5$), F ($-3,3$), G ($3,3$). Now, using straight lines, connect points A-B, B-C, C-D, D-E, E-F, F-G, G-B, G-D.
7. Compute the coordinates for the points on the drawing shown in Figure 5-16. The distance between each mark represents one inch.

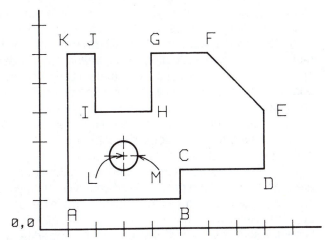

FIGURE 5-16 Compute the coordinates for each point on this drawing.

8. List some of the input devices used with an on screen menu.
9. List some of the cursor control devices used with function boards.
10. Describe the method used for selecting drawing functions on your CAD system.
11. In preparation for the next chapter, answer the following questions related to the CAD system you are learning:
    a. Identify the method used to select menu commands such as on-screen menu, pull-down menu, menu on tablet, and so forth.
    b. Identify the input device used to make menu selection.
    c. List all the different methods that can be used to make menu selection.
    d. Identify the location of any special keys used with the software.
    e. What type of marker is used to identify digitized points?
    f. Idenfity the location of any prompts or message lines.

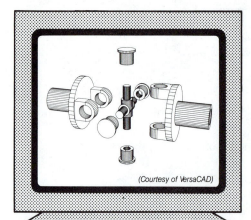

(Courtesy of VersaCAD)

SECTION 3

CREATING CAD DRAWINGS

Chapter Six

Drafting Fundamentals

This chapter introduces the reader to the basics of drawing using CAD. The chapter is divided into three major sections. The first section covers such basics of a CAD system as the menu, keyboard, and screen cursor; how to start up a system; creating drawing files; using grids; menu items; and the basic drawing functions of points and erase. The second part of the chapter covers the drawing of lines. The final section of the chapter shows how to draw a simple object and how to change line types and thicknesses using CAD.

The method used in this text to teach the fundamentals of CAD, given the diversity of CAD systems on the market, has been to describe the sequence of steps in general terms. As described here, the creating of CAD drawings begins by using a menu for command selection. A *command* can be defined as a specific word or phrase used to actuate a CAD system to perform a task. This chapter starts with menu command POINT. All menu items covered in this chapter are listed in the order in which they are covered, along with a brief definition. Menu items are printed in capital letters in the steps describing a drawing function to make it easy to identify a menu item when it is used.

All the necessary background material has been covered, and this and the next three chapters will help you to experience the numerous drawing functions available on most CAD systems. Before beginning these chapters, make sure that you know the four main parts of your system, understand Cartesian and polar coordinates, and have reviewed the operator's manual on the basics of your system.

PREREQUISITES

Before starting on this unit of instruction, you should be able to:

Identify

- the components of the CAD system.
- the method used to select commands on your CAD system.
- the device used to make menu selection.
- the cursor control device on the CAD system.
- the location of any special keys used for CAD.
- the location of prompts or message lines.
- the location of the screen cursor and control it.

Demonstrate

- a knowledge of basic drafting techniques, including orthographic projection and the alphabet of lines.

Use

- the coordinate system and polar system of measurement to place points.

OBJECTIVES

After completing this chapter, you will be able to

- start up the CAD system and ready it for drawing.
- create a file for drawings.
- control and use a grid for simple drawing tasks.
- draw lines, place points, and draw simple objects.
- use the erase command to delete various drawing entities.
- change the line type.
- change line thickness.
- create a 3-view drawing of a simple object.
- assign pen numbers to drawing entities.
- make a hard copy of the drawings on screen.
- plot a drawing

MENU ITEMS IN THIS CHAPTER

1. POINT—command used to define a location necessary to locate drawing entities.
2. GRID ON—command used to place a grid on screen.
3. GRID OFF—command used to turn a screen grid off.
4. ERASE—command used to erase entities singly or in groups, sometimes called DELETE.
5. ERASE WINDOW—command used to erase entities within a defined AREA or WINDOW.
6. ERASE ALL—command used to erase all displayed entities.
7. REVOKE—command used to delete the last entity or command implemented.
8. RECALL ENTITY—command used to bring back the last entity or group of entities that was erased or revoked, sometimes called OOPS or UNDELETE.
9. CLIP—command used to erase part of an entity, sometimes called TRIM or BREAK.
10. REDRAW—command used to redraw, repaint, or refresh the drawing image on screen.
11. HELP—command used to offer assistance to the operator in the form of prompts or messages.

12. SAVE—command used to store or create a file for a drawing so that it may be recalled at a later time. May be called **FILE**.
13. QUIT—command used to terminate a drawing, sometimes called **EXIT** or **END**.
14. RETURN—a command that must be executed after an input statement to cause action by the computer.
15. TYPED INPUT—command used to tell the computer that you would like to use the keyboard to input values. The operator will use this command to avoid using the menu to input values.
16. XYPT—command used for point selection using X-Y coordinates.
17. POLAR—command used for point selection using polar coordinates.
18. RESET ORIGIN—command used to change the location of the origin on the screen.
19. INPUT—command used to alert the computer to the fact that values will be input using the menu instead of from the keyboard.
20. CNST LN—command used to draw construction lines on the screen to assist in laying out a drawing. These lines are usually not part of the final plot.
21. HORZ—command used to draw horizontal lines or construction lines.
22. VERT—command used to draw vertical lines or construction lines.
23. DXDY—command using X-Y coordinates that are referenced from the last referenced point on the screen instead of from the origin.
24. LINE—command used to draw lines on the screen.
25. MOVE—command used to pick up the computer's pen and move to a new starting point to draw separate lines.
26. CHAIN—command used to place construction lines parallel to and at a distance referenced from the last placed construction line.
27. DATUM—command used to place construction lines parallel to and at a distance referenced from a base construction line.
28. LINE TYPE—command used to change the line type being used by the system.
29. LINE WEIGHT—command used to change the line thickness used by the system.
30. PRINT—command used to produce a hard copy of the screen image.
31. PEN NUMBER—command used to control the pen number assigned to various entities for CAD systems that have multicolored terminals or plotters.
32. SCALE—command used to change the scale of the drawing on screen.
33. PLOT—command used to PLOT drawing.

BASICS—LEARNING TO USE THE SYSTEM

Before attempting to draw using CAD you must have a firm grasp of the basics of your system. This is necessary to avoid confusion and frustration when learning on CAD. To learn the basics of a system it is necessary to read over carefully the manual provided with your CAD system to gain an understanding of the method used to create a drawing. Use that manual to determine what type of command selection and command-picking device is used, the method used to store and retrieve drawings, and how to start up the system, thus creating a general feeling of confidence and ease with the system. After each basic function is explained in this text, read the similar section in your manual to get a fundamental understanding of how your system handles that function.

THE KEYBOARD

The keyboard is a common feature on virtually all CAD systems and is used as an input device. The keyboard is a standard typewriter keyboard with additional special keys used for drafting and programming. Sometimes, a numeric keypad is included for mathematical calculations, menu and cursor control, and numeric input for labels. The keyboard's primary function is to interact with the system and for alphanumeric input. Look carefully at your keyboard and determine if there are any special keys and what their function may be. The manual provided with the system usually shows the keyboard and identifies the major areas and functions.

THE MENU- AND COMMAND-PICKING DEVICE

The next task that must be performed by the operator is to identify the method used to make command selection and the device used to pick the commands. Described next are the most common methods used to select command options.

On Screen Menu

Command options appear on screen and are activated by an input device . Light pens can be used for on screen menu selection by pointing the light pen at the desired command to activate the task. Another method is to use a puck or a mouse with a tablet to control the screen cursor. The input device is moved on the tablet to position the cursor over the command option on screen and a button is depressed on the input device to activate the menu option. Important parts of a screen are identified in Figure 6-1.

Menu Placed on a Graphics Tablet

In this example, a paper menu is placed over the graphics tablet and two corners are located by the input device to alert the computer to the position of the menu on the tablet. Common menu-picking devices are the stylus or the puck. The tip of the stylus or the cross hairs of the puck are located over the menu item, and the command option is selected by depressing a button on the input device or by pressing the tip of the stylus. Also, these picking devices are usually the method used to control cursor location on screen. This is accomplished by leaving an area on the tablet blank or free from menu options. Moving the input device over the blank area on the tablet causes movement of the cursor on the screen.

Programmed Function Keyboard

This method of command selection involves the use of a special keyboard with each button corresponding to a specific command. Depressing a button activates a command, and usually results in a sub-menu appearing on screen. Sometimes there are lights under the buttons to alert the operator to the functions that are active at that time. The cursor control device can be a joystick, a stylus or puck and tablet, a mouse, or a light pen.

Making the Choice

After reading through this chapter, your manual, and earlier chapters, you should be able to determine the type of menu and command-picking device that you will be using. Before continuing, you should determine: 1) the method of making command choices, i.e., menu on tablet, menu on screen, or programmed function board; 2) the menu-picking device; and 3) the cursor control device. Determining these factors is a good start in understanding the operation of your CAD system.

THE SCREEN CURSOR

The screen cursor is controlled by the input device or keyboard input. When drawing, the cursor is usually a small cross hairs or plus sign (+). See Figure 6-1. The location of the cursor determines end points of lines, centers of holes, entities to be erased, and numerous other drawing functions.

Many systems will have an area of the screen that will report the x-y coordinate position of the cursor. See Figure 6.1. On many systems, the cursor will change from a plus sign to a short line or dash (—). This usually occurs when keyboard input is necessary. For example, when drawing a circle of a specified diameter, the plus sign cursor might be positioned by the operator where the center of the circle is to be located. After the center is located, the plus sign cursor might automatically change to the short line cursor along with a prompt asking for the diameter. This indicates that the system needs keyboard input and that the screen cursor control device is inactive. The diameter of the circle is keyed-in and the return key pressed, causing the circle to be drawn and the plus sign cursor to return to the screen.

It is important for the operator to know what device is used for cursor control, what it looks like ordinarily, and what it looks like when keyboard input is necessary.

SYSTEM PROMPTS

Most CAD systems use a method of interacting with the user to make operation as easy as possible. One of the most common methods of interaction is through the use of prompts. *Prompts,* which are located on the CRT or on a separate alphanumeric terminal, help the operator to complete a specific task. For example, if an operator wants to specify a certain type of crosshatching pattern for a sectional drawing, the system will use prompts to help the operator input the correct information in the proper sequence. The system might first prompt the user to input the desired angle of the crosshatching. Then the system might prompt the user to indicate the spacing between the lines. It is important for the operator to be familiar with the location of prompts on the system and how to respond to them. Prompts are usually located at the top or bottom of the screen or on a separate alphanumeric terminal for dual CRT CAD systems.

CAD SYSTEM START-UP PROCEDURES

The steps to follow in starting up a CAD system vary widely from system to system. Described here are the general procedures to follow in starting up a standalone system, and a terminal tied into a mainframe or minicomputer. REFER TO THE MANUAL PROVIDED WITH YOUR SYSTEM TO FOLLOW THE SPECIFIC STEPS NECESSARY FOR START-UP.

Starting Up a Standalone System

The steps to follow in starting up a standalone CAD system should be similar to the ones described here.

1. Turn on the peripheral devices, such as the plotter, disk drive, tablet, and so forth. The switches can be located by referring to the manual.
2. Insert the software needed for drafting if it is not automatically booted (loaded and self-starting) into the system when the computer is turned on or located on the hard disk. This software is usually on a floppy disk or cassette tape for stand-alone systems.
3. Turn on the computer or terminal. The drafting software is usually booted into the system at this time. The operator responds to any prompts that may occur, such as inputting a date.
4. Insert the memory medium used to save any programs or drawings to be created. The memory medium is usually a floppy disk.

5. After the software is loaded, the menu will become functional, offering the operator access to choose the first command necessary to begin drawing. The operator does this through the input device or the keyboard.

6. The operator may now be asked to give a name to the drawing to be created, and to choose the drawing format, such as 17" x 11". The operator may also key-in the name of a previously created drawing to open an old file. The length of the name used is usually limited to a certain number of characters which varies from system to system. To open an old file, it is important that the name be keyed-in exactly as it was when saving the drawing. This means that spacing, periods, spelling, and so forth, must be just as they were when the drawing was saved. After this has been done, the operator is free to start drawing.

7. Terminating drawings on the CAD system usually involves the same steps as turning the system on, except that the steps are reversed. Of course, the operator should make sure that the drawing has been properly saved to avoid loss of data.

Logging-On a Mainframe System

Turning on a mainframe system is usually referred to as *logging-on*. The steps described here are only in general terms, and will vary from system to system.

1. The terminal is turned on and the system software is automatically loaded from memory.

2. The system will now automatically prompt for information necessary to run the software. The first prompt is usually for input of the operator's name, followed by pressing the return key. This is referred to as *logging-on*.

3. A password must now be input so that only authorized personnel may use the system. Passwords allow operators access to their private drawing files and prevent other users from having access to their files. In CAD, a *file* can be defined as the memory location for data created or input into the computer, which is then stored or saved on disk or tape in a specific memory location.

4. A short description of the drafting software will usually appear on screen. This is followed by a prompt or screen menu allowing the operator to create a new file or to open an old file. To create a new file a name has to be given to the drawing. The sheet size and units to be used may have to be determined at this time. The name is keyed-in to the computer, creating a file in memory to store the drawing. To open an existing file, the user must key-in the name used when it was created. This automatically retrieves the drawing and displays it on screen. Whether starting a new drawing or revising an old drawing, the system should be ready to begin drawing.

5. Logging-off the terminal involves terminating the drawing activities and exiting the drafting software. The operator usually has to just key-in

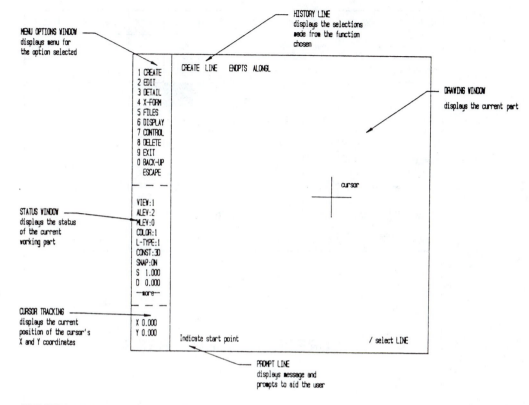

FIGURE 6-1 Different areas of a CAD display. *(Courtesy of CADKEY)*

"log-off" or "LO" to close the file. The system usually displays the date and time, and the name of the person who logged-off. The terminal power can then be turned off.

Default Values and Installing the CAD Software

After the CAD software has been loaded there are a number of items that are preset by the software that the user may want to change. For example, the default dimensioning system might be fractional dimensions; you may want to use 2-place decimal inches or metric measurements. That default value would have to be changed by the user before dimensions are placed on the drawing. One common method used by CAD systems to determine default values is through a *configuration* or *install* program. This program will have a series of prompts or menus that the user can select and change to control default values. Common default values that can be controlled by the user include colors, grid and snap spacing, lettering height and style, dimension type, metric or English units, size of chamfers and fillets, and so forth. These defaults will remain constant until changed by the user through the configuration program or through menu commands such as SET or CONTROL.

HELP	SCALE	SPLINE	LINE	HORZ	VERT	MOVE	ARC	CIRCLE	FILLET	SCALE	EDIT
RESAVE	MIRROR	FULL SCREEN	TANG	PARAL	PERP	ANGLE	C&R	C&D	CHAMFER	DATUM	CHAIN
REDRAW	COPY	DRAG	SERIES	TOLER	GEO TOL	SAVE SYMBOL	3 PTS	2 PTS	DXDY	CNST LN	INPUT
SAVE	DUPL	GROUP	MULT	DIM HGT	MEAS DIST	CALL SYMBOL	■	SELECT LAYER	(	)	/
QUIT	ZOOM	BLANK	JUSTIFY	UNITS	MEAS ANGLE	SAVE PARAM		CHANGE LAYER	*	−	+
RETURN	UNZOOM	UNBLANK	ANGLE	ANGULAR	CALC AREA	DIGITIZE		DISPLAY LAYER	,	.	0
TYPED INPUT	PAN	TRANS-LATE	SLANT	RADIAL	LEADERS	FILES		LINE TYPE	1	2	3
XYPT	GRID OFF	PLOT	HGT	DIA	RECT	ELLIPSE	DEFINE	LINE WEIGHT	4	5	6
POLAR	GRID ON	PRINT	TEXT	DIMEN	SLOT	HOLE	HATCH	PEN NUMBER	7	8	9
ERASE	ERASE ALL	ERASE AREA	CLIP	REVOKE	RECALL ENTITY	ROTATE	STRETCH	X	Y	A	L

CURSOR CONTROL AREA

FIGURE 6-2 Menu used in this text showing the commands and the cursor control area. This type of menu would be located on a graphics or digitizer tablet.

The configuration or install program is usually run after the CAD software is loaded onto the hard disk or the first time the software is used. This program is usually menu-driven with a list of menu items used to select the type of input devices, output devices such as printers and plotters, graphics display devices, and other options. This is necessary because many CAD software programs will run on a wide variety of hardware devices. The software must be installed or configured for your particular workstation before it will run successfully.

THE ERASE OR DELETE COMMAND

The erase command is one of the most important functions that a beginner can learn. Making errors is a common occurrence when learning CAD. For this reason, knowing how to access the erase command and how it works is very important to the new operator. The erase command may be called "delete," or something similar, but ERASE is used in this text, and it is one of the menu items shown in Figure 6-2. Figure 6-2 shows a listing of many of the menu commands used in this text, as well as modifiers (+, −, *, for example) and numbers.

Within the ERASE command, many methods are used to erase entities or groups of entities. Once the ERASE command is selected, the entity or unit to be erased is then digitized or selected from the menu. For example, referring to Figure 6-2 and Figure 6-3, a line can be erased from the screen by using the following steps.

1. Select the menu item labeled ERASE.
2. Select the item to be erased from the menu. In this example, LINE will be selected.
3. Move the screen cursor over the line to be erased and digitize that point.
4. The line will be deleted from the screen or a marker will identify the entity selected and a prompt will usually come on screen asking the operator to verify that the proper line was deleted by the computer. The operator must respond by hitting the Y or RETURN key for yes or the N key for no. If N is selected, the line is drawn back on the screen. Some CAD systems have a REVOKE command that can be used to recall the last entity deleted, thus eliminating the need for a prompt. The REVOKE command is covered in more detail later in this chapter.

Erase a Defined Area of a Drawing

The ERASE command can be used to erase virtually anything created with the system. Items can be deleted singly, such as lines, circles, text, dimensions or crosshatching. Items can be deleted as a group, such as all lines, circles, and text, for example. Items can also be deleted from a defined area, such as the top view of a three-view drawing. Deleting an area is accomplished by selecting menu item ERASE AREA or ERASE WINDOW and digitizing two corners of a box that includes the area to be erased. Any entity lying completely within the defined box would be deleted, Figure 6-4.

Erase All

Groups of instructions can be erased by using menu item ERASE ALL. This command must be combined with the command of the entity to be erased. For example, to erase all lines, you would select menu item ERASE ALL followed by menu item LINE. Figure 6-4 and the following steps demonstrate how to erase a group of entities using ERASE ALL.

1. Select menu item ERASE ALL.
2. Select menu item CONST LN (construction lines).

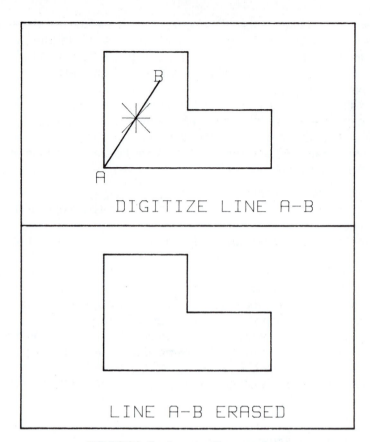

FIGURE 6-3 Erasing an entity on screen.

3. A prompt will usually appear asking to verify that you want to erase a group of entities in one command. The reason for this is to make certain that the operator did not mistakenly select the ERASE ALL command.

4. The user would respond yes or no to the prompt by keying-in or using the input device. If yes is selected on a raster terminal, all of the construction lines will be immediately deleted, as shown in the figure. On some terminals, a REDRAW or REFRESH of the screen may have to be executed before the construction lines are deleted from the screen.

Erasing Parts of Entities Using the Clip Command

Quite often, it is necessary to erase or *clip* only part of an entity, such as part of a line, an arc, or a circle. For example, when designing a part, you may decide that the length should be 5″ instead of 6″. Instead of erasing the line and redrawing a 5″ line, you can shield the line and selec-

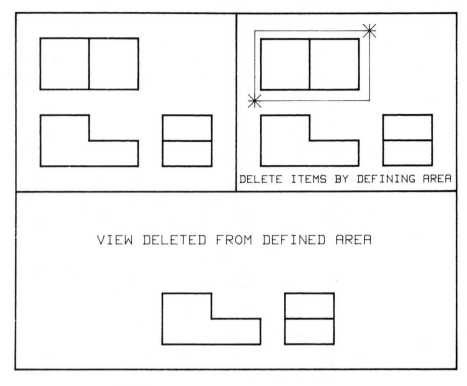

DELETE ITEMS BY DEFINING AREA

VIEW DELETED FROM DEFINED AREA

FIGURE 6-4 Deleting entities located in a defined area.

tively erase 1″ from the line. Standard drafting practices use a shield with small patterns cut into it to allow the selective erasure of entities. With CAD, a command can be used to clip entities as you would use a shield to erase entities on paper. The command used in this book is CLIP. Figure 6-6 and the following steps describe how the CLIP command is used with CAD.

1. Select menu item CLIP.
2. Select menu item LINE. This alerts the system that the entity to be clipped is a line.
3. Digitize the line to be clipped (1). This alerts the system to the specific line to be clipped.
4. Digitize two points along the line (2,3). The line segment between digitized points 2,3 will then be deleted as shown in the figure.

Revoking the Last Command or Entity Executed

Most CAD systems have the capability to revoke, delete or erase the last entity placed on the drawing or the last command executed. Menu item REVOKE ENTITY is used in this text to represent this command. The advantage of this command over the ERASE command is that the last entity

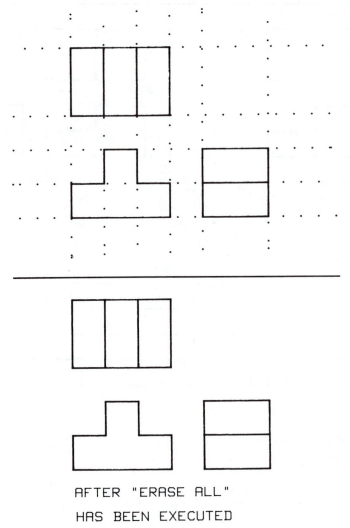

AFTER "ERASE ALL"

HAS BEEN EXECUTED

FIGURE 6-5 All construction lines deleted using ERASE ALL.

placed is automatically deleted without having to digitize the entity. Some CAD systems can "remember" the last 10 or more items added to the drawing and, by repeatedly selecting REVOKE ENTITY, the last 10 items can be deleted. Figure 6-7 and the following steps show how to use this command.

1. Select menu item REVOKE ENTITY. The last line entered will be deleted (1-2).
2. Select menu item REVOKE ENTITY. The line entered before line 1-2 will be deleted (2-3).
3. Select menu item REVOKE ENTITY. The line entered before line 2-3 will be deleted (2-4).

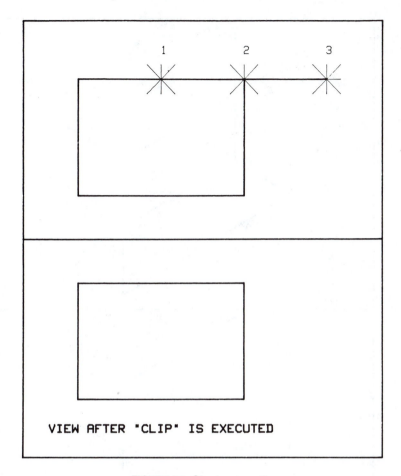

FIGURE 6-6 Clipping an entity.

Recalling the Last Entity or Command Erased

Most CAD systems have the capability to recall the last entity erased or revoked. Menu item RECALL ENTITY is used here to replace items that have been deleted. This command will automatically replace the last item deleted from the drawing. As with REVOKE ENTITY, some CAD systems can recall the last 10 or more items that have been erased. Figure 6-8 and the following steps demonstrate the RECALL ENTITY command.

1. Select menu item RECALL ENTITY. The last entity erased will be re-drawn (hole).
2. Select menu item RECALL ENTITY. The hidden line deleted before the hole will be redrawn (1-2).
3. Select menu item RECALL ENTITY. The hidden line deleted before line 1-2 will be redrawn (3-4).

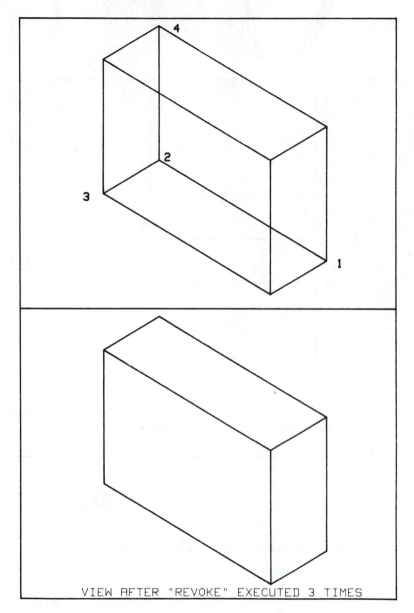

VIEW AFTER "REVOKE" EXECUTED 3 TIMES

FIGURE 6-7 Deleting entities using REVOKE.

CHANGING THE SIZE OR SCALE OF THE DRAWING SHOWN ON SCREEN

All CAD systems have the capability of *factoring* or *scaling* a drawing on screen. The scale of the drawing is usually reduced to allow more room for drawing or to add more details. Scaling can also be used to enlarge or shrink a drawing to fit different sizes of drawing media for a final

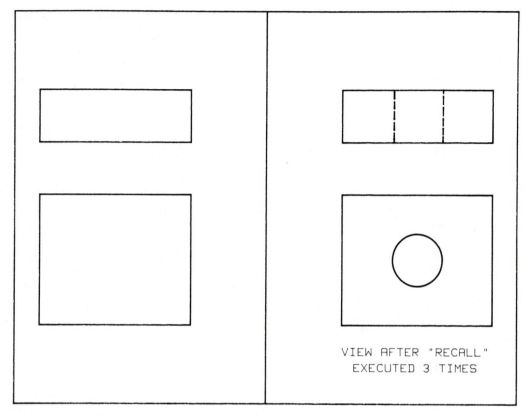

FIGURE 6-8 Recalling entities that had been previously erased.

plot. Menu item SCALE is used here to change the size of drawings. After the menu item SCALE is selected, the user must key-in the new scale for the drawing. The operator usually has the option of inputting different scales for the X and Y axes. Scales are input as ½ or .5 for half scale, 2/1 or 2 for double scale. Figure 6-9 and the following steps show how to change the scale of a drawing.

1. Select menu item SCALE.
2. A prompt will usually appear asking you to input the X-axis scale. The operator keys-in the desired scale (.75).
3. The prompt will return asking you to input the Y-axis scale. The operator keys-in the desired scale (.75). The computer will redraw the picture to the scale input, as shown in the figure.

AUDIBLE OUTPUT

Audible tones are often used by CAD systems to communicate with the operator. Some systems will use two types of audible tones: one long and one short. The short tone is used when an item is selected from the menu to alert the user that the computer has received the command. The long tone (or series of tones) is used to signify an operator's error. For example, if the operator wants to erase a circle and the point digitized is too far away from the circle, a long tone will sound and the circle will not be erased. Sometimes the long tone will be accompanied by a prompt explaining the problem to the user.

REFRESH OR REDRAW COMMAND

This command is used to repaint or redraw the current drawing on screen. This command is used with vector storage CRTs that do not have selective erasure to repaint the screen image whenever entities have been erased. This menu function is used with raster displays to redraw the current drawing if, for example, the grid was turned off or if the user wanted to remove marker. If the GRID OFF command was executed, a REDRAW command may also be executed for the grid to be taken off the screen. Figure 6-10 shows this example. This is only one of the examples of how the REDRAW command is used. You will find that this command is used quite often when drawing with CAD.

HELP COMMAND

The HELP command is a feature found on many CAD systems to provide assistance when drawing. For example, if the operator would like to erase a line and does not know the steps to follow, the HELP command could be selected and prompts would appear giving the operator the steps to follow to perform the task. This is an extremely helpful command for beginners, but an operator should not become dependent upon it as experience is gained. Dependency on the HELP command will cause the drafter to become slow and less productive.

MAKING HARD-COPY CHECK PRINTS

One feature or peripheral device included with many CAD systems is a hard-copy device. This device is used primarily for making a copy of the screen image that can be used by the designer as a check print. This print is sometimes referred to as a *screen dump*, because the image on screen is "dumped" into the printer line by line. The image produced by the hard-copy device is usually not to scale and is of poor quality, especially if taken from a raster CRT. Figure 6-11 shows a screen dump produced by a thermal copy printer.

To produce this rough copy or check print a menu function or special key on the terminal must be selected by the user. In this text, menu item PRINT is used to produce a rough screen print of a drawing. This PRINT function can be executed at any time by the operator.

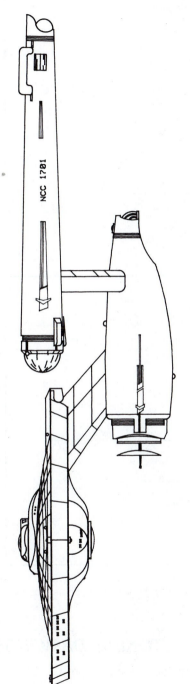

FULL SCALE DRAWING

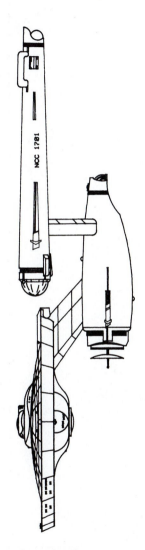

VIEW AFTER 3/4 "SCALE" IS EXECUTED

FIGURE 6-9 Scaling the screen image of a drawing.

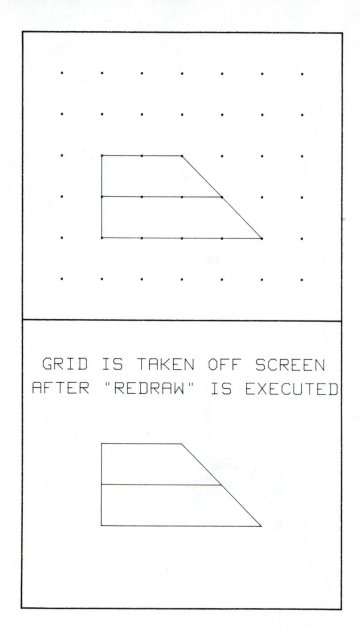

FIGURE 6-10 Refreshing or repainting the screen to update the image.

PREPARING A DISK FOR STORING DRAWINGS

After a drawing has been created, the data stored in RAM memory must be transferred to a permanent memory device to avoid loss of work when the terminal is turned off. If the memory device to be used is a floppy disk or microdisk, these disks must be initialized or formatted before data can be saved. This is done to new disks and it can be used to erase old disks of unwanted files. The *formatting* process organizes the surface of the disk into tracks or sectors so that the information stored goes into a predetermined place where it can be found later. The formatting process is usually a submenu item listed under "files" or the "files management" command. After the formatting process is executed, the disk drive may spend up to ten minutes formatting the disk.

SAVING A DRAWING

The SAVE, "file" or "stop" command is used when you have either completed a drawing or have finished working for that day. When the SAVE command is executed, the drawing is stored on the memory device for permanent storage or as a means of avoiding loss of work if the computer is turned off. Many systems allow the user to input or change the file name and other pertinent information when the SAVE command is executed. When this information has been input, the drawing will be saved.

Resaving a Drawing

Often, a design takes many days or weeks to complete. On some CAD systems, a separate file would be created if the SAVE command were to be used after each day's work. This would unnecessarily use up valuable memory storage space. To avoid this, a command such as RESAVE can be used. This command will update the old file instead of creating a separate new file for each day's work. Some systems will automatically update the old file as long as the same name is used when it is saved. Carefully review your system manual to determine the best procedure to follow in saving a drawing that will take a number of days to complete.

THE QUIT COMMAND

The QUIT, EXIT, END, or KILL command will terminate a drawing without saving it in memory. If the drawing had been previously saved, the file in memory would not be affected by executing QUIT. Only work done subsequent to saving the drawing would be lost. Operators should be sure of how to execute this command on their CAD system to prevent lost work.

THE RETURN COMMAND

The RETURN command is used to end an input statement. This command may be referred to as "carriage return," "execute," "continue" or "space bar," depending upon the CAD system being used. In addition, this

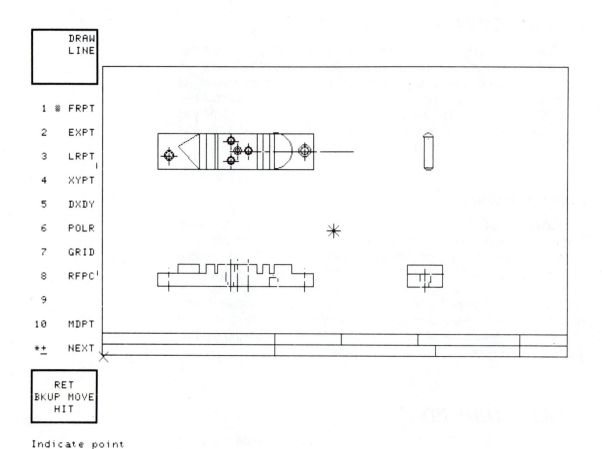

DRAW
LINE

1 ※ FRPT

2   EXPT

3   LRPT

4   XYPT

5   DXDY

6   POLR

7   GRID

8   RFPC

9

10   MDPT

*+   NEXT

RET
BKUP MOVE
HIT

Indicate point

PN01 LT03 LV01

FIGURE 6-11 A thermal copy print (screen dump) of the screen image.

command may be a menu item that is selected after keying-in the value. For example, drawing a line using X-Y coordinates will involve keying-in the X and Y values. However, the computer will not take action on these values until the RETURN command is executed. By using this command, you are telling the computer that you have completed your input and you want the computer to take action or execute your commands.

POINT CHOICE COMMANDS

A very important concept that must be understood by the operator is the "point choice command." A CAD system draws by referencing points. To draw a line from A to B, points A and B must be located in some way and those location points communicated to the computer. Failure to do this will result in inaccurate drawings. The method used to select points varies from system to system. Many CAD systems have more than one method of selecting point choices. The reason for reviewing the Cartesian coordinates and polar coordinates in Chapter Five was because of the large number of systems that use coordinates to identify points. Other point choice commands used include "grid points," "constructions lines," "unreferenced points," "last referenced point," "existing points," "center," "perpendicular," "tangent," "midpoint," and so forth.

Unreferenced Points

An *unreferenced point* is a point that is placed on the screen in a position without regard to other points on the screen. These points may also be called *free points* or *digitized points*. An unreferenced point is input into the computer by moving the cursor with the input device to the location desired and digitizing that point. This point is used for sketching a design or to begin a drawing. When starting a drawing, since there are no points on the screen to reference except the origin, a point has to be selected to start the drawing. This first point could also be selected by coordinates or by grid points. Figure 6-12 shows an example of lines drawn using unreferenced points.

The following steps show how unreferenced points are used. To do this, there must be a menu item for points or construction points. This menu item will be called POINT.

1. Select menu item POINT.
2. Move the screen cursor to the location desired for a point and digitize that point (A). A small dot or X will appear at the position that was digitized. This X is referred to as the *last referenced point* and shows the position of the imaginary pen on the screen.
3. Repeat step 2 to locate more points (B,C,D).
   Figure 6-13 shows how these points might be represented on a CAD system.

FIGURE 6-12 Lines drawn using unreferenced points.

Using Screen Grids and SNAP To Facilitate Drawing

One method of creating a drawing with CAD is to use a screen grid. A *grid* is a series of small dots arranged in rows across the screen. These dots are arranged in rows horizontally and vertically creating a pattern on screen that can be used as reference points to draw entities and locate details for a drawing. Some systems also allow the rows of dots to be rotated outward producing a radial grid. (A rectangular grid made from a series of dots is shown in Figure 6-10.) Grids are not considered part of the drawing and will not plot.

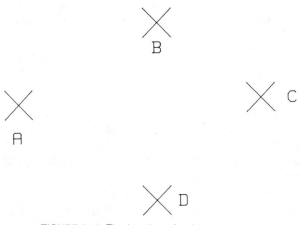

FIGURE 6-13 The location of points on a screen.

The spacing of the dots is controlled by the operator through keyboard input. The command used to activate the grid is a menu item or special key on the keyboard. This textbook uses the GRID ON command to activate a screen grid. GRID OFF is used to turn the screen grid off. Grid points are used as reference points for drawing such entities as lines, circles, and so forth. To use a screen grid, the cursor is located close to the grid point using the input device. When the cursor is in position on the screen, that point is digitized causing a reference point to "snap to" or locate on the nearest digitized grid point. The operator does not have to be located exactly on the desired grid point because the system will search and find the nearest grid point to the digitized spot and locate on the grid. This SNAP feature is a menu command that can be turned on and off independent of the grid. The SNAP spacing can be changed just as the grid spacing can be changed.

Turning On and Using Grid Points

To use a grid, the screen grid must be turned on by the operator. A grid is created by selecting menu item GRID ON. After the menu item is selected, the operator must key-in the desired spacing of the grid points. For example, a grid with a spacing of 1″ is produced by keying in 1, or a grid of 1/2″ is made by keying-in .5. After the grid is on screen, the operator positions the screen cursor at the grid point desired, and then digitizes that location. The point choice is then referenced to the nearest grid point from the cursor. The following are the steps to use for placing a grid on the screen and for locating a point referenced to the grid.

1. Select menu item GRID ON and SNAP ON.
2. A prompt or a flashing cursor will appear indicating keyboard input for the spacing of the grid points. Input the spacing desired by keying-in .25 for grid points that are spaced .25″ apart followed by RETURN. A grid represented by small dots spaced .25″ apart horizontally and vertically will be drawn on the screen. An example of referencing points to screen grids is shown in Figure 6-14.
3. Menu item POINT is selected by the user.
4. The operator now positions the cursor in the location desired for a point (A) and digitizes that point. A point will be located at the nearest grid point to the cursor location when digitized.
5. More points can be selected by moving the cursor to the positions desired (B,C) and digitizing them.

Selecting Points Using X-Y Coordinates

The use of X-Y coordinates is one of the most popular methods used to make point selections with CAD. A basic understanding of coordinate geometry is required before using this method. The coordinates necessary to locate points can be input in two ways. One method is to use typed

FIGURE 6-14 Referencing points to screen grids.

input from the keyboard and menu item XYPT. This is a menu command that is labeled TYPED INPUT. Another is to use the menu for assigning coordinate values.

When using typed input, the screen cursor may not be displayed. Instead, a flashing cursor is displayed near the border of the screen or on the alphanumeric display terminal. Using Cartesian coordinates, the operator would select menu item XYPT and key-in the X and Y values, such as X2,Y4, followed by RETURN or "carriage return" key. This specifies a point 2 units in the X direction from the origin, and 4 units in the Y direction from the origin. It is important that the operator inputs information into the computer in the proper format. If the computer does not receive information in the proper form, a tone will sound and a message such as "syntax error" or "illegal character" will be displayed.

Mathematical functions can also be included when defining points. Some of these functions are: $+$ addition, $-$ subtraction, $*$ multiplication, $/$ division. For example, a point can be defined using Cartesian coordinates in this manner: $X(6.375 - 4.125)*2.25$, $Y4.75/3*1.5$, which are equal to X5.0625, Y2.375. The computer will automatically perform the calculations and display the point. This saves the user time by letting the computer calculate the X-Y coordinates in the same step used to display the point.

When using coordinates to define points it is important that the operator know the position of the origin (0,0) on the system. Most systems have origins in the center of the screen or in the bottom left corner, Figure

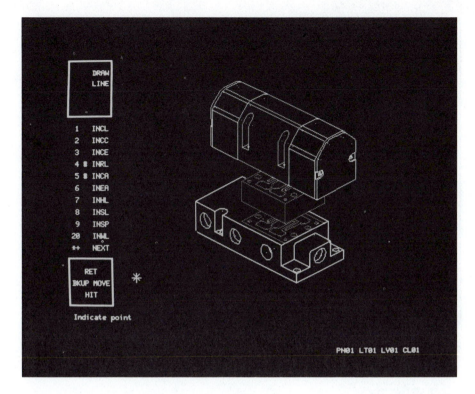

FIGURE 6-15 The origin is the asterisk-like symbol shown in the lower left corner of the screen. Notice also the menu and prompt line showing the words INDICATE POINT. *(Courtesy of Bruning CAD)*

6-15. However, it is possible to reset the origin to the user's specifications by using a menu command labeled RESET ORIGIN. When this command is executed, the operator must key-in the new position for the origin by keying-in values from the current origin's position. For example, Figure 6-16 shows the origin's original position (bottom left) and the new position. The new position was found by keying-in X5,Y5.

The following steps show how to display points using X-Y coordinates. Refer to Figure 6-17.

1. Select menu item POINT.
2. Select menu item XYPT.
3. Select menu item TYPED INPUT.
4. Key-in X2,Y2 and press the RETURN key.
5. A point will be displayed on the screen.
6. Key-in X4,Y4 RETURN and X6,Y3 RETURN.
7. Points will be displayed on the screen as shown in the figure.

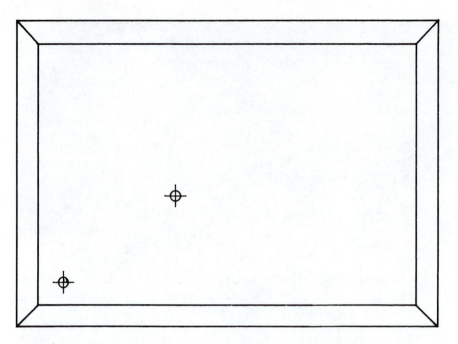

FIGURE 6-16 Resetting the position of the origin on screen.

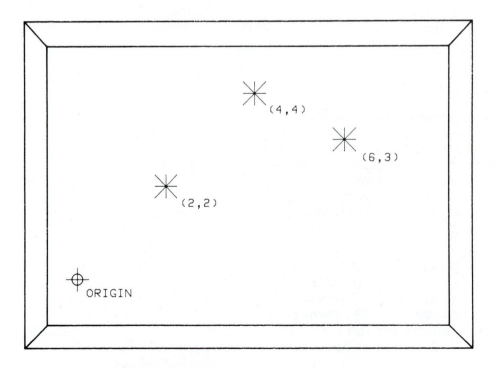

FIGURE 6-17 Locating points on screen using X-Y coordinates.

Using the Menu for Coordinates

The previous examples show coordinates being assigned by keyboard input. The menu may also be used to assign coordinate values for points on many CAD systems. This is accomplished by assigning numbers and letters to an area of the menu. The menu in Figure 6-2 shows the numbers and modifiers located in the lower right corner. The procedures to follow to place points are slightly different from those used with keyboard input; additional menu items must be added. An INPUT command is used to alert the computer to the fact that values will be input using the menu. The RETURN command is used to alert the computer that the input is complete. The numbers Ø through 9, the decimal point and comma, the letters X,Y,A,L, and the mathematical functions +, −, *, /, ( ) are used to assign coordinate values.

The following example shows how coordinates can be used to locate points using the menu for inputting values.

1. Select menu item POINT.
2. Select menu item INPUT.
3. Select menu item X4 COMMA Y4 RETURN.
4. A point will be placed on the screen 4 units along the X axis and 4 units along the Y axis.

Selecting Points Using Polar Coordinates

Points can also be selected by keying-in the polar coordinates of the position by using the menu command POLAR. Two values must be used to identify points using polar coordinates, followed by a carriage return. A typical polar coordinate would be A45,L2. The A value is the angle from the origin and the L value is the length from the origin measured along the angle. These values would locate a point 45 degrees and 2 units from the origin. Each succeeding point would be referenced from the last point input if in the incremental or relative coordinate mode. Location 45,2 would become the new origin or the point that the next polar coordinate is referenced from. Polar and Cartesian coordinates can be mixed on most systems. For example, A9Ø,Y4 or XØ,L4 would represent the same point.

The following steps show how to place points using polar coordinates. Refer to Figure 6-18.

1. Select menu item POINT.
2. Select menu item POLAR.
3. Select menu item TYPED INPUT.
4. Key-in the values A9Ø,L4 followed by pressing the RETURN key.
5. A point will be displayed on the screen.
6. Key-in A − 3Ø,L3 and RETURN. This point would be referenced from the last point input (A9Ø,L4).
7. Key-in A45,L4 and RETURN. This point would be referenced from the last point input (A − 3Ø,L3). These points will be displayed as shown in the figure.

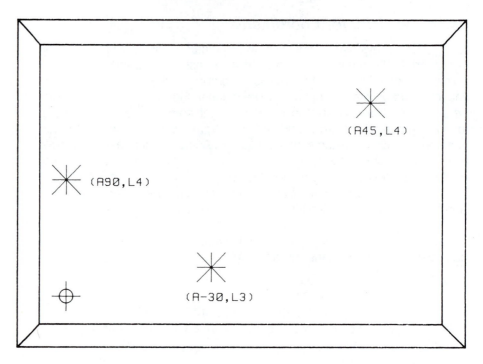

FIGURE 6-18 Locating points on screen using polar coordinates.

Using Drawing Entities To Select Points

Locating points can also be accomplished by referencing to drawing entities that are already on the screen. *Entities* are drawing features such as lines, circles, arcs, and splines. As a drawing is being made, points can be used from drawing features such as the ends of lines or the intersection of two lines. This eliminates the need to key-in coordinates or to reference to a grid point.

Figure 6-19 shows examples of the following drawing features that can be referenced to for point or entity location.

1. Incremental or the X-Y distance from the last referenced point (LRPT). The LRPT is usually represented on a screen as an X.
2. Intersecting construction lines.
3. The center of such entities as circles and arcs.
4. Ends of such entities as lines, circles, arcs, and splines.
5. The intersection of two entities such as two lines, two circles, or two arcs; a line and an entity; an arc and an entity; a circle and an entity; and a construction line and an entity.

Using drawing features to place points is done by selecting the menu item POINT and moving the cursor to the item and digitizing the point. A point will be placed on the item at the position digitized.

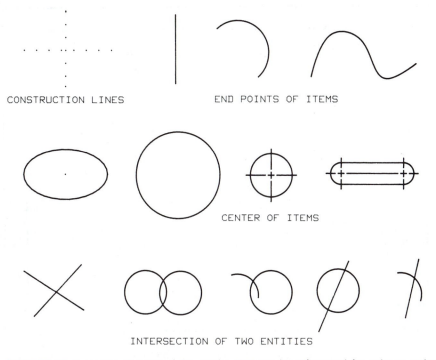

CONSTRUCTION LINES

END POINTS OF ITEMS

CENTER OF ITEMS

INTERSECTION OF TWO ENTITIES

FIGURE 6-19 Examples of some of the entities that can be referenced for point or entity location.

Using Construction Lines To Place Points

Another method used to select points is to place construction lines on the screen and to use the points of intersecting construction lines as positions for points. This procedure is very similar to that used by a drafter on a board. The drafter usually places light construction lines on the drawing before darkening-in the details. Some CAD systems function in a similar way, by placing construction lines on the screen and then referencing to these lines to place the drawing details on the screen. These construction lines sometimes appear as lines made up of a series of dots or else as solid lines. Either way, they can usually be left on a drawing and the computer will disregard them when making the final plot or they may be deleted before plotting.

Steps Used To Draw Construction Lines

Using construction lines for point choice is done by selecting POINT on the menu and moving the cursor to the intersecting point of the construction lines and digitizing that position. A point will be placed at the intersecting point selected.

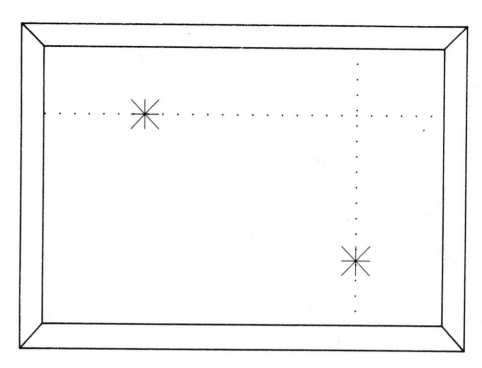

FIGURE 6-20 Drawing construction lines.

1. Select menu item CNST LN (construction line).
2. Select menu item HORZ, to draw horizontal construction lines.
3. Move the cursor to a location on the screen to lay out a construction line and digitize that point. A horizontal construction line will be drawn on the screen through the cursor location when digitized. See Figure 6-20.
4. Select menu item VERT to draw vertical construction lines.
5. Move the cursor to the position desired for the vertical construction line. Digitize that point to produce a vertical construction line drawn through the cursor location. The intersecting point of the construction lines can be used as a reference point for locating other points or drawing entities.

Incremental or a Distance from the Last Referenced Point

Points can be placed by referencing to the last point input. For example, if the last point input was X4,Y4, using menu command DXDY a new point can be placed on the screen referenced from 4,4 instead of from the origin (Ø,Ø). The following example shows the steps that would be used for incremental point selection. Refer to Figure 6-21.

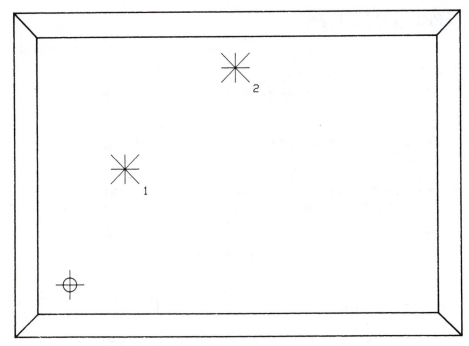

FIGURE 6-21 Locating points from the last referenced point.

1. Select menu item POINT.
2. Select menu item XYPT.
3. Select menu item TYPED INPUT.
4. Key-in X2,Y4 and RETURN.
5. Select menu item DXDY.
6. Key-in DX4,DY3.5 and RETURN.
7. Two points will be displayed on the screen, the second one referenced from the first, by using the DXDY command.

DRAWING LINES WITH A CAD SYSTEM

After the operator learns to select and place points using CAD, the drawing of lines can begin. Lines are drawn by locating the two end points of the line. These end points can be located by any of the methods used to locate points as described previously. After learning how to draw lines, the first simple object will be drawn by using the skills mastered in this chapter. The method used to select different line types is then covered so that more complicated objects can be drawn. Methods used to create 3-view drawings is the last section covered in this chapter. You will progress from drawing lines to the making of simple orthographic drawings using CAD.

PREREQUISITES

Before starting on this unit of instruction, you should be able to:

Locate

- points on the CAD system using referenced or unreferenced points.

Place

- a grid of any specified size on the screen and place points using that grid.
- points using X-Y and polar coordinates.
- construction lines on the screen and reference points to those lines.
- points using drawing entities previously located on the screen.

Produce

- a hard-copy print of a drawing on screen.

OBJECTIVES

After completing this section of the chapter, you will be able to

- draw lines using X-Y and polar coordinates.
- draw lines from referenced or unreferenced points.
- draw lines using construction lines for point choice.
- draw lines using grid points.

DRAWING LINES USING UNREFERENCED POINTS AND REFERENCED POINTS

Drawing lines using unreferenced points is similar to drawing free-hand or without instruments on paper. The menu function LINE is used to draw object lines. The menu item LINE is selected and the cursor is positioned where the line is to start and that point is digitized. A small dot or X will appear at this point on most CAD systems. This X is referred to as the *last referenced point,* and is the location of the pen on the screen. The cursor is then positioned where the line is to end and that point is digitized. A line will be drawn between the two digitized points on the screen.

The following steps show how a line is drawn using unreferenced points. Refer to Figure 6-22.

1. Select menu item LINE.
2. Move the cursor to the position for the start of the line and digitize that point (A). An X should appear on the screen representing your first digitized point.
3. Move the cursor to the desired end point of the line and digitize that point (B). A line will now be drawn from point A to point B. Point B now

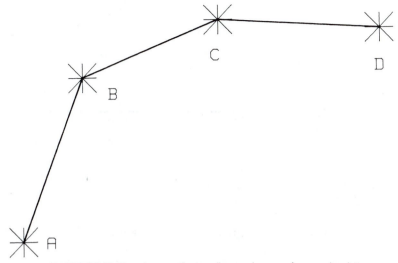

FIGURE 6-22 Drawing continuous lines using unreferenced points.

becomes the last referenced point and usually has an X marker at the end point of the line. To draw a continuous line it is necessary to digitize more points on the screen, causing lines to be drawn from the last referenced point or the end point of the last line drawn to the new digitized points. Points C and D are digitized creating lines BC and CD in the figure.

THE MOVE COMMAND

To draw a series of separate lines it is necessary to use another command from the menu to alert the computer to "pick up its pen" and move the starting point of another line to the next digitized point without drawing a line between the points. This will create a space from the last referenced point to the starting point of a separate line. The menu function used to make the computer pick up its pen may be a "comma" or "semi-colon" or such words as "separate line" or "move." This text uses the menu command MOVE to make the computer pick up its pen.

The steps to follow in drawing separate lines are shown as follows. Refer to Figure 6-23.

1. Select menu item LINE.
2. Move the cursor to the position for the start of the line and digitize that point (A).
3. Move the cursor to the desired end point of the line and digitize that point (B). A line will be drawn from point A to point B.
4. Select menu item MOVE.
5. Digitize the starting point for the next line (C). A small X will appear on the screen at the digitized point.
6. Select menu item LINE.

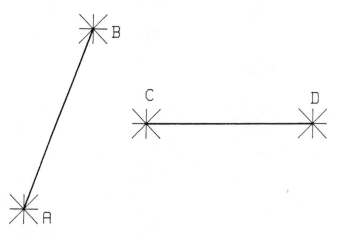

FIGURE 6-23 Drawing separate lines using unreferenced points.

7. Digitize the ending point for the line (D). A line will be drawn between points C and D, creating two separate lines. Repeating steps 4 through 7 will allow you to fill the screen with separate lines.

USING GRID POINTS FOR DRAWING LINES

Using grid points to reference the ends of lines allows the operator to draw lines that are horizontal, vertical and of a known length, unlike those drawn using unreferenced points. To draw lines using a grid, it is necessary to place a grid on the screen and digitize two grid points. A line will be drawn between the two digitized grid points.

The following steps describe how grid points are used to draw lines. Refer to Figure 6-24.

1. Select menu item GRID ON.
2. Answer the prompt by keying-in the space desired between grid points, followed by RETURN. A grid will now fill the screen.
3. Select menu item LINE.
4. Move the cursor nearest the grid point desired for the start of the line and digitize that point (A). An X will be shown on the screen on the grid point nearest the cursor location when digitized.
5. Move the cursor to the desired end point of the line and digitize that point (B). A line will now be drawn between the grid points that were digitized (AB).

CREATING A SIMPLE DRAWING USING GRID

To create a simple drawing using GRID it is necessary to use the MOVE command and to digitize grid points. The following steps show the operator how to create a simple drawing, and demonstrate one of the great-

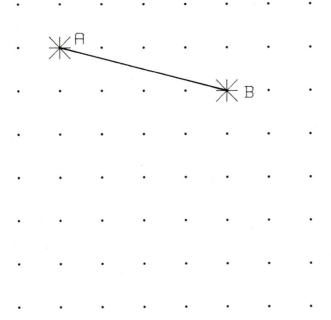

FIGURE 6-24 Drawing lines by referencing to grid points.

est advantages of using CAD. Once an operator gains experience, the speed and accuracy of drawing this simple object could not be matched using traditional drafting methods. Refer to Figure 6-25.

1. Select menu item GRID ON.
2. Key-in a value for the space desired to complete the drawing followed by RETURN. The grid spacing in this example is 1".
3. Select menu item LINE.
4. Move the cursor to the nearest grid point for the start of the view and digitize that point (1).
5. The first line is 4". Move the cursor 4 grid points horizontally to the right and digitize that point (2). A 4" horizontal line will be drawn across the screen.
6. The next line is 2" drawn vertically from the last referenced point. Move the cursor 2 grid points vertically from point 2 and digitize that point (3). A line will be drawn vertically from point 2 to point 3.
7. A line must now be drawn 4" to the left of point 3. Move the cursor 4 grid points to the left and digitize that point (4). A 4" horizontal line will now be drawn from point 3 to point 4.
8. A 2" vertical line is drawn from point 4 back to point 1. Move the cursor 2 grid points vertically down and digitize point 1. A 2" vertical line is now drawn on screen completing a 2" x 4" rectangle.
9. To complete the view, it is necessary to draw diagonal lines between the corners of the rectangle. This is most effectively accomplished by

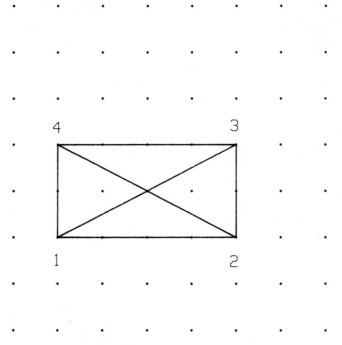

FIGURE 6-25 Creating a drawing by referencing to grid points.

moving the cursor to point 3 and digitizing that point. A line would then be drawn from point 1 to point 3.

10. Select menu item MOVE.
11. Move the cursor to point 2 and digitize that point. This now becomes the last referenced point.
12. Select menu item LINE.
13. Move the cursor to point 4 and digitize it. A line will be drawn from point 2 to point 4 completing the view.
14. Use PRINT for check-copy.

The Importance of Using the Move Command

An important concept was demonstrated in steps 10 and 11 that may not be apparent to the beginning operator. The reason for the MOVE command was to pick up the pen and move it from point 3 to point 2. This was done so that another line was not drawn over the top of line 2-3. If the operator had simply stayed in the LINE command and moved the cursor from point 3 to point 2 and digitized that point for the start of diagonal line 2-4, another line would have been drawn over the top of line 2-3. The operator would not be able to tell that two lines had been drawn until the final plot of the object was done. At that time, the plotter would draw line 2-3 twice on the paper. For this simple drawing, the double line would have little effect on the final plot, except that line 2-3 might look a little thicker than the other lines on the drawing.

The major problem associated with creating double lines occurs in more complicated drawings, because double lines will unnecessarily slow down the speed of the plot. When moving the last referenced point from one position to another, remember to use the MOVE command to avoid double lines. Some systems will use a command other than MOVE to accomplish this same task. For example, a "comma" or "semicolon" can be used or a "line" command that draws separate lines may be used instead of MOVE.

DRAWING LINES USING X-Y COORDINATES

The procedures used here are similar to those explained earlier in this chapter for locating points, except that menu item LINE is used in place of the item POINTS. Menu item TYPED INPUT is also used to key-in the X-Y coordinates for the view.

Before starting a drawing using coordinates, it is necessary to locate the position of the origin (Ø,Ø). The default position is usually located in the center of the screen or the bottom left corner of the screen. *Default* is the word used to describe the value assigned to certain functions when the computer is turned on. The origin can be moved by using a menu item such as RESET ORIGIN.

The origin's position is usually represented on screen by a +, Ø, or an ∗. This textbook uses Ø to represent the location of the origin. In this example, the origin will be moved from the default position in the lower left corner of the screen to the lower left corner of the view to be drawn. See Figure 6-26.

1. Select menu item RESET ORIGIN.
2. Select menu item TYPED INPUT.
3. Key-in X2,Y2 and RETURN. This will move the origin to position labeled 1 in the figure and be represented by a Ø.
4. Select menu item LINE.
5. Select menu item XYPT.
6. Key-in X4,YØ and RETURN, and a line will be drawn from point 1 to point 2.
7. Key-in X4,Y4 and RETURN, and line 2-3 will be drawn.
8. Key-in X2,Y4 and RETURN, and line 3-4 will be drawn.
9. Key-in X2,Y2 and RETURN, and line 4-5 will be drawn.
10. Key-in XØ,Y2 and RETURN, and line 5-6 will be drawn.
11. Key-in XØ,YØ and RETURN, and line 6-1 will be drawn, completing the view.
12. Select menu item PRINT for a check-print.

Using X-Y Coordinates Referenced from the Last Input Point

The previous example uses the origin as a reference to locate new points. Another method of using X-Y coordinates is to input X-Y values referenced from the last point input. On some CAD systems, this is automatically

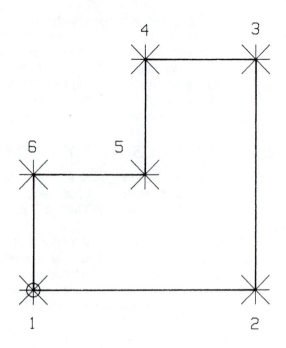

OLD ORIGIN

FIGURE 6-26 Creating a drawing using X-Y coordinates.

done using XYPT; other systems reference from the origin when using XYPT. For this reason, another command must be available in order to use X-Y coordinates referencing from the last point input. This procedure uses the menu command DXDY to locate a new point using coordinates from the last referenced point. **Refer to Figure 6-27.**

1. Select menu item MOVE.
2. Select menu item XYPT.
3. Key-in X2,Y2 and RETURN. This will place the pen at position 2,2 (1) on the screen.
4. Select menu item LINE.
5. Select menu item DXDY.
6. Key-in DX4,DYØ and RETURN to draw a line from 1-2.
7. Key-in DXØ,DY1.5 and RETURN to draw line 2-3.
8. Key-in DX-2,DYØ and RETURN to draw line 3-4.
9. Key-in DX-1,DY1.5 and RETURN to draw line 4-5.
10. Key-in DX-1,DYØ and RETURN to draw line 5-6.
11. Key-in DXØ,DY-3 and RETURN to draw line 6-1.
12. Select menu item PRINT to make a check-print.

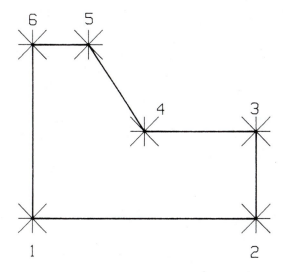

FIGURE 6-27 Creating a drawing using X-Y coordinates referenced from the last point input.

As you can see, the last referenced point becomes a new Ø,Ø position and, to get the next point, X-Y coordinates must be referenced from the last referenced point. However, the origin in this example is still in its default position of the lower left corner. Keying-in a D before the X and Y tells the computer to reference the new X,Y position from the last referenced point.

THE USE OF POLAR COORDINATES TO DRAW LINES

Polar coordinates can also be used to draw lines and views by using the letter A for the angle of the line in degrees and the letter L for the length of the line referenced from the origin. The units for L do not have to be specified, as the number will take on the value of the units currently being used. For example, if working in inches, a 4.75 input after L would be in inches; if working in millimeters, the number 50 would be in millimeters. Figure 6-28 will be drawn using polar coordinates.

1. Select menu item MOVE.
2. Key-in A45,L2 and RETURN to move the last referenced point to position 1 from the origin.
3. Select menu item LINE.
4. Select menu item POLAR.

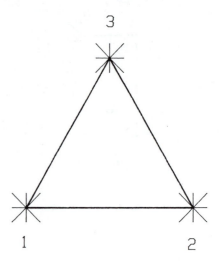

FIGURE 6-28 Creating a drawing using polar coordinates.

5. Key-in AØ,L3 and RETURN to draw a horizontal line of 3″ length (1-2).
6. Key-in A12Ø,L3 and RETURN to draw line 2-3.
7. Key-in A24Ø,L3 and RETURN to draw line 3-1 to complete the view.

DRAWING LINES BETWEEN EXISTING POINTS

As a drawing is being produced on a CAD system, lines can be drawn on the view without referencing to grid points or X-Y or polar coordinates. This is accomplished by referencing the end points of lines to be drawn to existing points on a drawing. An *existing point* is a point located on drawing entities, such as the end points of lines, the center of circles, and the ends of an arc. (Figure 6-19 shows examples of existing points located on drawing features.)

The uncompleted drawing on the left in Figure 6-29 must have lines drawn between existing points 1-2 and 2-3. This is accomplished by using the menu function LINE, and moving the cursor to the existing points and digitizing them. The computer will "snap to" or locate the nearest existing point to the cursor location when digitized, allowing a line to be drawn by referencing to existing drawing features.

1. Select menu item MOVE.
2. Move the cursor to existing point 1 and digitize it.
3. Select menu item LINE.
4. Move the cursor to existing point 2 and digitize it. A line will be drawn from points 1 to 2.

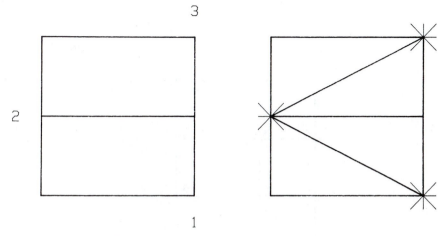

FIGURE 6-29 Drawing lines from existing points.

5. Move the cursor to existing point 3 and digitize it. Line 2-3 will be drawn, completing the view shown on the right side of the figure.

USING CONSTRUCTION LINES TO DRAW LINES

Construction lines in drafting are very light lines that are placed on paper to lay out a drawing. Construction lines in CAD are used for the same reason as they are in traditional drafting. These lines usually appear as very short dashed lines, or as lines made up of small dots as shown in Figure 6-30. The following steps show how construction lines might be created and used to draw a view.

1. Select menu item CNST LN (construction line).
2. Select menu item HORZ (horizontal).
3. Select menu item CHAIN. This menu item uses the last drawn construction line as a reference from which to place succeeding construction lines.
4. Move the cursor to a location on the screen to lay out a construction line for the base of the view and digitize that point (A). A construction line will be drawn horizontally across the screen through the cursor location when digitized.
5. A prompt will ask to input the distance to the next horizontal construction line. Key-in the required distance of 1 and press RETURN. Another horizontal construction line will be drawn 1″ from the first line (B).
6. The prompt will appear again. Key-in 1 and press RETURN. Construction line (C) will be drawn.
7. Key-in 1 and RETURN to answer the prompt and place the last horizontal construction line (D).

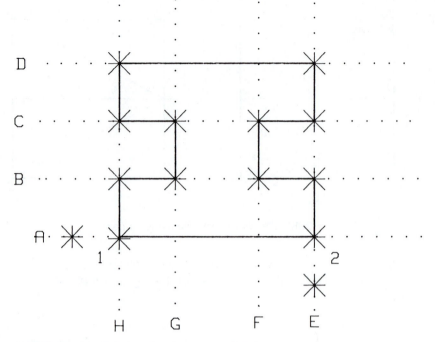

FIGURE 6-30 Using construction lines as reference points to create a drawing.

8. Select menu item VERT (vertical construction line).
9. Select menu item DATUM. DATUM uses the first construction line placed as the reference for each succeeding construction line. CHAIN could have been used, but DATUM is used to demonstrate how this command functions.
10. Move the cursor to the position desired for the first vertical construction line. This could be either to the right or the left of the screen. In this example, the right side of the screen is chosen. Digitize that point to produce construction line E.
11. Respond to the prompt by keying-in −1 and RETURN to draw construction line F. A negative value is necessary because moving to the left of the datum line is a negative value on an X-Y coordinate. Recall also that moving to the right of an X-Y coordinate is positive, moving downward is negative, and moving upward is positive.
12. Key-in −2.5 and RETURN to draw construction line G.
13. Key-in −3.5 and RETURN to draw construction line H.
14. Select menu item MOVE.
15. Move the cursor to point 1 and digitize that point. The intersecting point of the construction lines is an existing point that the pen "snaps to" for the starting point of the line.
16. Select menu item LINE.
17. Move the cursor to point 2 and digitize it to produce line 1-2.

All the succeeding lines are produced in the same way that line 1-2 was produced in step 17, by moving counterclockwise around the perimeter of the view and digitizing each corner to produce the final drawing.

CREATING A SIMPLE ORTHOGRAPHIC DRAWING USING CAD

The ability to select points and draw lines will enable the operator to create simple 3-view drawings of an object. This section covers the sequence of steps used to draw a 3-view drawing from start-up of a CAD system to saving the final drawing. How to change line types and line thicknesses is also covered. Three different methods of drawing orthographic views are presented: grid, X-Y coordinates, and construction lines.

PREREQUISITES

Before starting on this unit of instruction, you should be able to:

Locate

- points and lines using a grid, X-Y coordinates, or construction lines.

Draw

- lines using grid, X-Y coordinates, polar coordinates, or construction lines.

Use

- the MOVE command.

OBJECTIVES

After completing this section of the chapter, you will be able to

- change line types on the CAD system.
- change the line thickness on the CAD system.
- draw simple orthographic views using CAD.

CHANGING THE LINE TYPE

The *alphabet of lines* was created so that a standard of different line types could be used to represent objects more clearly. For example, a *hidden line* is used to represent edges, surfaces or features located behind another drawing feature or surface. Hidden lines are represented by a series of dashed lines equally spaced. After having used a CAD system to

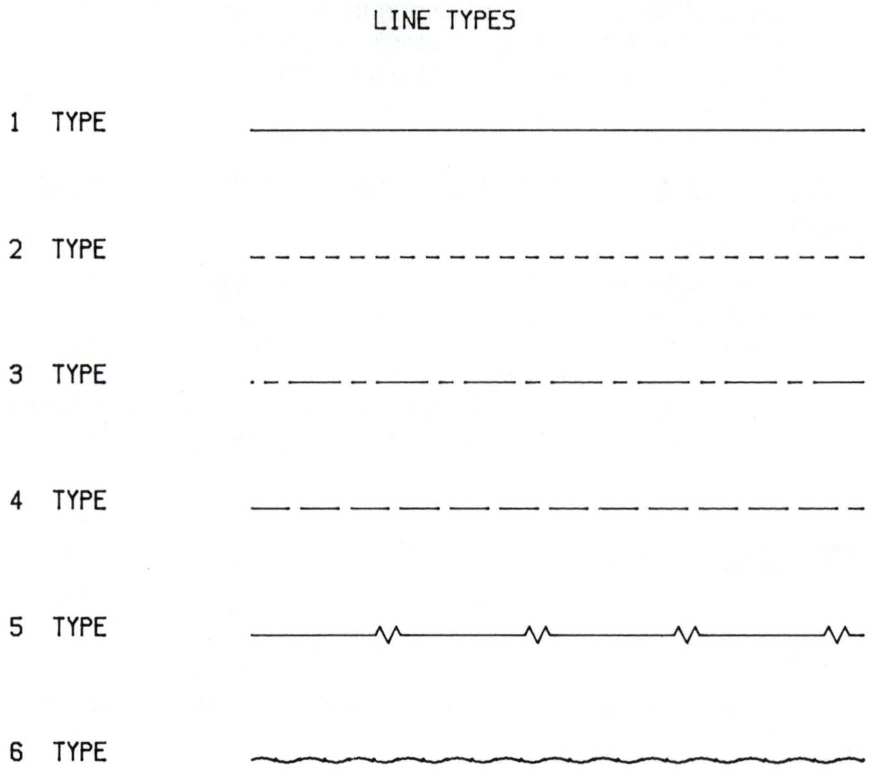

1 TYPE _____

2 TYPE –

3 TYPE . – —— – —— – —— – —— – —— – —— – —— – —

4 TYPE —— —— —— —— —— —— —— —— —— —— —

5 TYPE ————/\/————/\/————/\/————/\/

6 TYPE ～～～～～～～～～～～～～～～～

FIGURE 6-31 Common examples of the various line types used with CAD.

draw lines, it should be apparent that solid lines are the *default lines* of the system.

To create a drawing with different line types it is necessary to change the default line type. This is accomplished by using a menu item to access the different line types available, and to change the line type to be drawn on screen. This menu item is selected before the line is drawn. The types and number of lines available to the user vary on each system. However, all systems have all the line types needed to complete most mechanical and architectural drawings, including solid, dashed, center, and phantom lines. Some systems allow users to create their own lines and make them part of the menu choices. Some systems also allow the user to choose different line weights in addition to different line types.

The following steps change line types using menu function LINE TYPE. Figure 6-31 shows the six different line types used in this text. They are selected by choosing the number assigned to each line type (as shown on the menu in Figure 6-2).

1. Select menu item LINE TYPE.
2. Select menu item 2. Menu item 2 will change the line type to a dashed line.
3. Select menu item LINE.

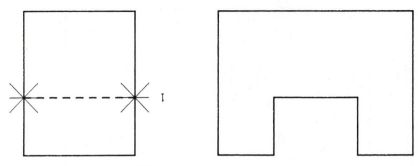

FIGURE 6-32 Drawing a hidden line with CAD.

4. Move the cursor to position 1 and digitize it.
5. Move to position 2 and digitize it. A dashed line will be drawn from the last referenced point to the newly digitized point, as shown in Figure 6-32. The system will continue drawing dashed lines until menu item LINE TYPE is chosen and the line number changed.

CHANGING THE LINE WEIGHT

To change the line weight it is necessary to assign numbers to different line weights, 1 being the thinnest and 6 being thickest for purposes of this text. The following steps show how to change line weights. Figure 6-33 shows squares drawn at different line weights.

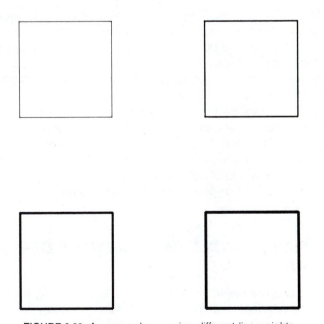

FIGURE 6-33 A square drawn using different line weights.

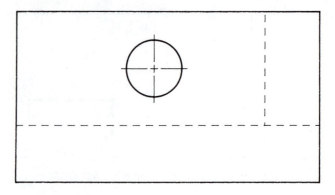

FIGURE 6-34 / An example drawing using two different line weights.

1. Select menu item LINE WEIGHT.
2. Select menu item 6 to draw a thick line.
3. Select menu item LINE.
4. Position the cursor on the screen and digitize that point. A thick line will be drawn on screen from the last referenced point to the digitized point.

      Figure 6-34 shows an object drawn using two different line weights. The object lines are thick, and the center lines and hidden lines are thin.

Changing Pen Numbers

      Many CAD systems have pen plotters that allow the user to control the number of the pen assigned to each drawing entity. This enables the user to control the thickness of lines inasmuch as pens are available in many different thicknesses. It also allows the user to produce multicolored plots. Multicolored plots are very useful for improving legibility, especially for architectural and electronic drawings.

      Most pen plotters have between one and eight pens. The user can control the pen number assigned to each entity through a menu command or a special key. Changing the pen numbers of entities must be done before the entity is drawn. Menu item PEN NUMBER is used in this text for changing pen numbers assigned to entities. This command may also be used to control the color of entities displayed on color CRT terminals.

DEVELOPING A SIMPLE ORTHOGRAPHIC DRAWING USING CAD

      All the prerequisites for creating a simple drawing have now been covered. The operator should have the skills necessary to make orthographic drawings. The first complete drawing to be created will be of the

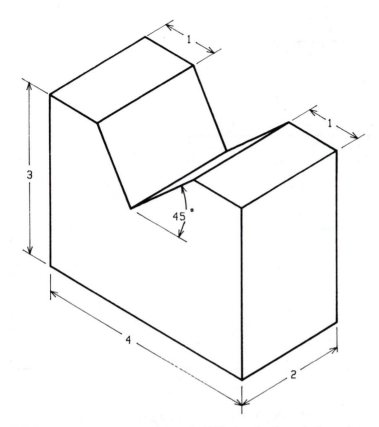

FIGURE 6-35 Dimensioned drawing of a V-Block to be drawn in three views.

V-Block shown in Figure 6-35. A 3-view drawing of this object will be created using a grid. The object will then be drawn using X-Y coordinates and, finally, using construction lines.

DRAWING THE V-BLOCK USING GRID POINTS

For the relatively simple V-Block drawing, using grid points would be a quick method of producing 3 views. Whenever a drawing must be done on a CAD system, much time can be saved by thinking through the steps that might be needed to complete the task.

Looking at the drawing, you can quickly determine that the grid spacing could be as large as 1″ because the smallest dimension of the V-Block is 1″. The side view should be the last view drawn because there is a hidden line that needs to be drawn. The line type will have to be changed, so this step is saved until last to prevent having to change the line back to solid, thus saving time. The amount of space used between the views should also be determined before starting. Because there are few details to be dimensioned, 2″ between the views will be sufficient. Refer to Figure 6-36 when reading through the following steps.

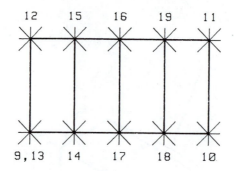

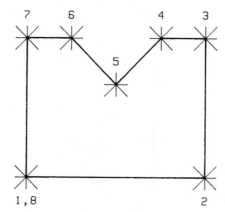

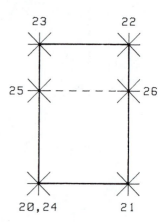

FIGURE 6-36 Drawing the V-Block using grid points.

1. Turn the CAD system on and load the software for drafting.
2. Create a file for the drawing.
3. Enter the drafting module.
4. Select menu item GRID ON.
5. Answer the prompt by keying-in 1 and RETURN for a 1″ screen grid.

Drawing the Front View

1. Select menu item LINE.
2. Move the cursor to the position desired on the screen to begin drawing the front view. This example starts with the lower left corner of the front view (1). Digitize this point.
3. Complete the view by moving the cursor counterclockwise around the perimeter of the view and successively digitize grid points 2 through 8.

Drawing the Top View

1. Select menu item MOVE.
2. Move the cursor 2 grid points vertically from point 7 and digitize. This will cause the last referenced point to move from point 1 to point 9.

3. Select menu item LINE.
4. Complete the outline of the view by moving the cursor counterclockwise around the perimeter of the view and successively digitize points 10 through 13.
5. Select menu item MOVE.
6. Move the cursor to grid point 14 and digitize.
7. Select menu item LINE.
8. Move the cursor to grid point 15 and digitize. Line 14-15 will be drawn.
9. Select menu item MOVE.
10. Move the cursor to grid point 16 and digitize.
11. Select menu item LINE.
12. Move the cursor to grid point 17 and digitize. Line 16-17 will be drawn.
13. Select menu item MOVE.
14. Move the cursor to grid point 18 and digitize.
15. Select menu item LINE.
16. Move the cursor to grid point 19 and digitize. Line 18-19 will be drawn, completing the top view.

Drawing the Side View

1. Select menu item MOVE.
2. Move the cursor to grid point 20, which is 2 grid points to the right of grid point 2, and digitize. This will allow a 2" space between the views.
3. Select menu item LINE.
4. Move the cursor to grid point 21 through grid point 24, digitizing each corner along the way. The perimeter of the side view will be drawn.
5. Select menu item LINE TYPE.
6. Select menu item 2 to change the line type to short dashes for hidden lines.
7. Select menu item MOVE.
8. Move the cursor to grid point 25 and digitize.
9. Select menu item LINE.
10. Move the cursor to grid point 26 and digitize. Hidden line 25-26 will be drawn, completing the side view and the orthographic drawing of the V-Block using grid points.
11. Use menu item PRINT for a hard copy.

Saving the Drawing

1. Select menu item SAVE.
2. Key-in the title of the drawing, if that was not already done when the file was created when starting the drawing. The drawing will be saved and can be recalled later to make changes or to add other features to it, such as notes.

DRAWING THE V-BLOCK USING X-Y COORDINATES

After studying the object, it is decided to select 2″ between the views and to draw the side view last because of the hidden line. Another consideration is where to set the origin to most effectively complete the drawing. Positioning the origin on the lower left corner on the front view is probably the best method to follow. All X-Y values will then be positive, simplifying the drawing of the object. Refer to Figure 6-37 when reading the following steps.

1. Turn on the CAD system and load the software for drafting.
2. Create a file for the drawing.
3. Enter the drafting module.
4. Select menu item RESET ORIGIN.
5. Move the cursor to a position on the screen near the lower left corner from which all views may be drawn, and digitize that location. The Ø sign will be located at the position of the new origin.

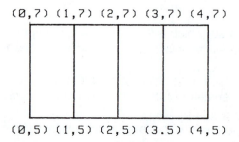

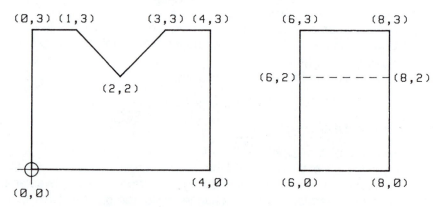

FIGURE 6-37 Drawing the V-Block using X-Y coordinates.

Drawing the Front View

1. Select menu item LINE.
2. Select menu item XYPT.
3. Key-in X4,Y0 and RETURN. A line will be drawn from 0,0 to 4,0.
4. Individually key-in the coordinates for each corner, going counter-clockwise around the perimeter of the view ending with 0,0. This completes the front view of the V-Block. Refer to Figure 6-37 for the coordinates for each corner.

Drawing the Top View

1. Select menu item MOVE.
2. Key-in X0,Y5 and RETURN to locate the lower left-hand corner of the top view 2″ from the front view.
3. Select menu item LINE.
4. Key-in X4,Y5 and RETURN to draw the line from 0,5 to 4,5.
5. Key-in X4,Y7 and RETURN, X0,Y7 and RETURN, and X0,Y5 and RETURN to draw the perimeter of the top view.
6. Select menu item MOVE.
7. Key-in X1,Y5 and RETURN to move the last referenced point without drawing a line.
8. Select menu item LINE. Key-in X1,Y7 and RETURN to draw line 1,5-1,7.
9. Select menu item MOVE.
10. Key-in X2,Y7 and RETURN.
11. Select menu item LINE.
12. Key-in X2,Y5 and RETURN to draw line 2,7-2,5.
13. Select menu item MOVE.
14. Key-in X3,Y5 and RETURN.
15. Select menu item LINE.
16. Key-in X3,Y7 and RETURN to complete the top view.

Drawing the Side View

1. Select menu item MOVE.
2. Key in X6,Y0 and RETURN to begin the right-side view 2″ from the front view.
3. Select menu item LINE.
4. Key-in X8,Y0 and RETURN to draw line 6,0-8,0.
5. Key-in X8,Y3 and RETURN to draw line 8,0-8,3.
6. Key-in X6,Y3 and RETURN to draw line 8,3-6,3.
7. Key-in X6,Y0 and RETURN to draw line 6,3-6,0. This will complete the perimeter of the side view.
8. Select menu item LINE TYPE.
9. Select menu item 2 to change the line type to a dashed line for drawing hidden lines.
10. Select menu item MOVE.

11. Key-in X6,Y2 and RETURN.
12. Select menu item LINE.
13. Key-in X8,Y2 and RETURN. The hidden line will be drawn, completing the V-Block using X-Y coordinates.
14. Use menu item PRINT for a hard copy.

Saving the Drawing

1. Select menu item SAVE.
2. Key-in the title of the drawing and other information, if this was not already done when the file was created to start the drawing. Press RETURN and the drawing will be saved to be recalled later for changes or to add other features to it, such as dimensions.

DRAWING THE V-BLOCK USING CONSTRUCTION LINES

Construction lines can be used to lay out the drawing using the intersecting points of the construction lines as reference points for digitizing the end points of lines. In this example, horizontal lines are placed on the screen followed by vertical construction lines. The CHAIN command is used to place construction lines for all horizontal and vertical details, surfaces and corners. *Chain* means to place a construction line at a distance referenced from the previously placed construction line. It is important to study the drawing before starting to determine the number of horizontal and vertical construction lines needed to complete the drawing. Every surface must have a construction line, and every corner must have a horizontal and a vertical construction line.

1. Turn the CAD system on and load the software for drafting.
2. Create a file for the drawing.
3. Enter the drafting module.

Placing Horizontal Construction Lines

1. Select menu item CNST LN.
2. Select menu item HORZ.
3. Select menu item CHAIN.
4. Move the cursor to a location on the screen near the bottom to locate the baseline for the front view, leaving enough room to draw the top view. Digitize this point (A) in Figure 6-38. A horizontal construction line will be drawn across the screen through the point where the cursor was located when digitized.
5. A prompt will usually appear asking the operator to input the distance to the next construction line. Key-in 2 and RETURN. A horizontal construction line (B) will be drawn 2″ from line A.

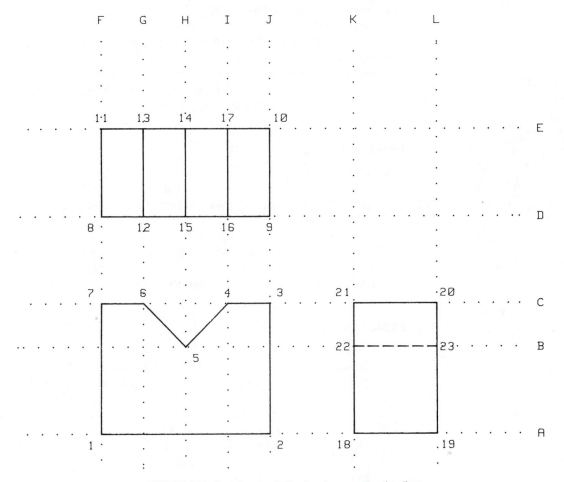

FIGURE 6-38 Drawing the V-Block using construction lines.

6. The prompt will appear again. Key-in 1 and RETURN to draw construction line C 1″ from line B.
7. Key-in 2 after the prompt appears and RETURN to draw construction line D 2″ from line C.
8. Key-in 2 and RETURN to place the last horizontal construction line (E) 2″ from construction line D.

Drawing Vertical Construction Lines

1. Select menu item VERT.
2. Select menu item CHAIN.
3. Move the cursor to the far left side of the screen and digitize. This will place the vertical construction line F that is to be used as the vertical baseline for the left end of the front and top view. The line was placed to the far left to leave room on screen for the right-side view.

4. Answer the prompt by keying-in 1 and RETURN. This will place the vertical construction line G 1″ from line F.
5. Key-in 1 and RETURN to draw construction line H.
6. Key-in 1 and RETURN to draw construction line I.
7. Key-in 1 and RETURN to draw construction line J.
8. Key-in 2 and RETURN to draw construction line K.
9. Key-in 2 and RETURN to draw construction line L.

Drawing the Front View

1. Select menu item MOVE.
2. Position the cursor at the intersection of construction lines A and F (point 1) and digitize to move the last referenced point.
3. Select menu item LINE.
4. Position the cursor at point 2 and digitize. Line 1-2 will be drawn.
5. Continue digitizing the intersecting construction line points, moving in a counterclockwise direction. Included are points 3 through 7 and back to point 1 to complete the front view.

Drawing the Top View

1. Select menu item MOVE.
2. Move the cursor to the intersection of construction lines D and F and digitize. This will move the last referenced point to point 8.
3. Select menu item LINE.
4. Digitize point 9 to draw line 8-9.
5. Digitize point 10 to draw line 9-10.
6. Digitize point 11 to draw line 10-11.
7. Digitize point 8 to draw line 11-8.
8. Select menu item MOVE.
9. Digitize point 12.
10. Select menu item LINE.
11. Digitize point 13 to draw line 12-13.
12. Select menu item MOVE.
13. Digitize point 14.
14. Select menu item LINE.
15. Digitize point 15 to draw line 14-15.
16. Select menu item MOVE.
17. Digitize point 16.
18. Select menu item LINE.
19. Digitize point 17 to draw line 16-17 which completes the top view.

Drawing the Right-side View

1. Select menu item MOVE.
2. Digitize point 18.
3. Select menu item LINE.

4. Digitize point 19 to draw line 18-19.
5. Digitize point 2Ø to draw line 19-2Ø.
6. Digitize point 21 to draw line 2Ø-21.
7. Digitize point 18 to draw line 21-18.
8. Select menu item LINE TYPE.
9. Select menu item 2 to change the line type from a solid line to a dashed line.
10. Select menu item MOVE.
11. Digitize point 22.
12. Select menu item LINE.
13. Digitize point 23 to draw hidden line 22-23 which completes the 3-view drawing of the V-Block using construction lines.
14. Use menu item PRINT for a hard copy.

PRINTING AND PLOTTING YOUR DESIGNS

In the process of creating a design or a part it is not uncommon to create a check-print or plot. Most CAD systems will give the user the option of creating output on paper through a dot-matrix printer, thermal printer, laser printer, or some other means of making a fast, rough copy of the design for a check print. This option is usually listed in the commands as PRINT. For high-quality final copies of the design a plot is usually made using a pen plotter, ink-jet printer, or electrostatic printer. The command used for creating a final plot is usually called PLOT. The difference between the two outputs is shown in Figures 6-39 and 6-40.

Most CAD systems will allow the user to make a print or plot at any time. Prints or plots can be made to any scale by inputting the desired size before the plot is made. Borders can be placed around the drawing and plotted on any size paper that the plotter is capable of using. It is also possible to enlarge parts of a drawing many times and make a check-plot or print of that particular area. This is shown in Figure 6-41. Plots can also be rotated and groups of entities or details can be left off the plot. Details can also be separated by level or layer and each layer or combination of layers plotted. For example, it is possible to separate details of an architectural floor plan by layers. Each layer could then be plotted independent of the other.

The plotter itself also offers many different variables for the user. Pen type, velocity, and pressure can all be controlled by the plotter hardware or CAD software. The drawing medium can also be controlled and varied. Most CAD software programs and plotters will offer the user a wide range of options when plotting.

CREATING A PLOTFILE AND MAKING A PLOT

The user has the option of making a print on paper at any time while making a drawing. As described earlier, a screen dump can be made using a hard-copy device such as a thermal printer. This is good for check prints

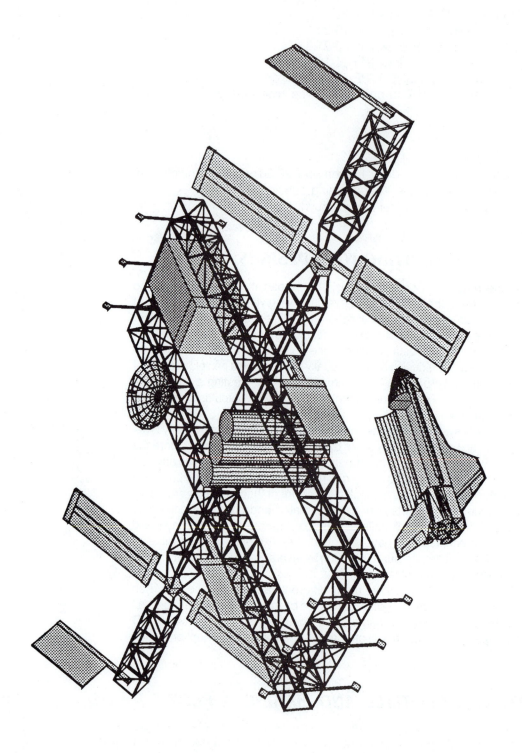

FIGURE 6-39 Laser printer-shaded output of a drawing. *(Courtesy of VersaCAD)*

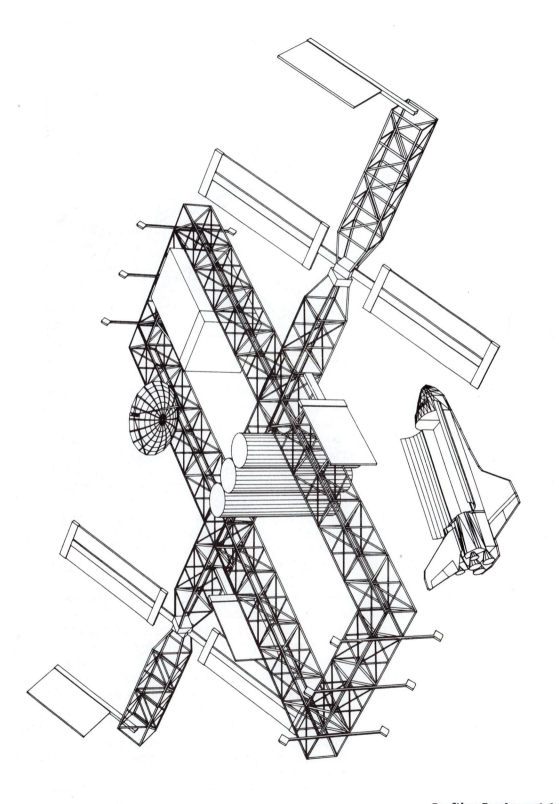

FIGURE 6-40 Plotter output of a drawing. *(Courtesy of VersaCAD)*

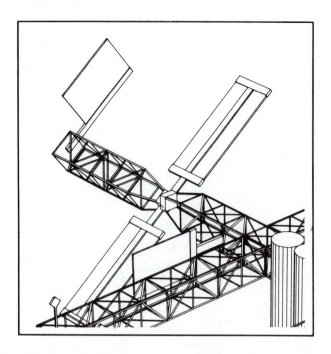

FIGURE 6-41 Plotter output of an enlarged area of a drawing. *(Courtesy of VersaCAD)*

or rough copies, but there are times when it is necessary to have a clean drawing produced on various drafting media such as Mylar®, vellum or paper. A plotter is used for reproducing part geometry on various drafting media. To produce a plot, it is necessary to create a plotfile which will convert the geometry into a form that can be used by a plotter.

With some CAD systems, the sheet size is selected and represented on screen before the drawing is started. For other systems, the drawing is made and then placed onto a sheet on screen after the drawing is produced. Regardless of how the drawing is created, once the drawing is completed and ready for plotting, a plotfile must be made. To create a plotfile, menu item PLOT is added to the commands. After selecting menu item PLOT, the user keys-in a name of the plotfile and a scale. The plotfile is then created by the computer. This file is then used to make a plot of the drawing by keying-in the parameters of the plot. Examples of parameters that can be controlled before plotting include the layers to be plotted, text on or off, and rotation of the part. Scaling is a very powerful option that allows the user to scale the drawing to virtually any scale after the drawing has been made. After the drawing is created, the user has the option of changing the scale at which it is to be plotted by keying-in the desired scale. This is a very convenient and time-saving operation. Because of the scaling ability before plotting, the user can create the drawing using full-scale dimensions. This eliminates the need to calculate scaled dimensions. With CAD, a drawing can be drawn at virtually any scale desired once the full-scale drawing has been produced. The user simply keys-in the desired scale and the plotter then makes the drawing to the desired scale.

Loading the Plotter

Before a plot is made, the plotter must be loaded with paper and turned on. For flatbed plotters, the paper is held electrostatically by pressing a "hold" button after the paper is positioned. For drum plotters, the paper is either fed from a roll, which does not require the loading of paper before a plot, or the paper is loaded and held in place by pinch rollers for single-sheet-fed drum plotters. For beltbed plotters, the paper is either fed from a roll or a sheet is taped onto the drum. Refer to the plotter manual for the exact sequence of steps necessary to follow for loading paper.

For some pen plotters, after the plotter is turned on it will automatically size the paper by moving the pen holder across the paper. After loading the paper on pen plotters, the user may have to change or place pens in the proper position on the plotter. When the plotter is ready, the user dumps the plotfile to the plotter. The plotter will take the plotfile and create a drawing on paper. Once the plot begins, the user usually has the option of pausing or stopping the plot completely by pressing the appropriate button on the plotter, keyboard or menu. After the plot is completed, the user unloads the paper and turns off the plotter. Because most plots take longer than a dump plot and are more expensive, the operator should make sure that the drawing is correct before plotting.

SUMMARY

Chapter Six introduced you to the basics of using a CAD system from starting up the system to drawing a simple object and saving a drawing. The three methods used to draw the V-Block are not the only methods that can be used to draw this object. Most CAD systems have so many diverse options and combinations of options that it is virtually impossible to describe them all in detail. As an operator, it is your responsibility to determine the most efficient method to develop a drawing. Your ability to choose the best way of drawing an object will improve with each drawing produced. At the end of this chapter, you will find some sample drawing exercises that will help you to develop your skills on a CAD system. Chapter Seven describes how to use basic geometry to draw more complex objects.

Chapter Six GLOSSARY

Command—a specific word or phrase used to provide the means for a CAD system to perform a task.

Default—a word used to describe the value assigned to certain functions when the computer is turned on. For example, the solid line is the default line used on a CAD system.

Entity—drawing features, such as lines, circles, arcs, and splines.

Existing point—a point located on drawing entities, such as the end points of lines, the center of circles, and the ends of an arc.

File—the memory location for data created or input into the computer.

Grid—a series of small dots arranged in rows on the screen that can be used for point selection.

Grid point—used as a reference point for drawing entities.

Last referenced point—the position of the pen on the screen, usually represented on screen with a small dot or an X.

Prompt—a message shown on a CRT to assist the operator in performing a task on a CAD system.

Unreferenced point—a point positioned on the screen without regard to other points on the screen.

Chapter Six REVIEW

1. Determine if your CAD system has any special keys on the keyboard programmed specifically for drafting.
2. Identify the method used to make command selections and the device used to pick those commands on your system.
3. Identify the location of the menu or programmed function board on your CAD system.
4. List the steps used to start-up your CAD system.
5. List the steps used to place a grid on screen.
6. List the steps to follow to delete an entity.
7. Explain the function of audible sound as used on your CAD system.
8. Explain how to use the HELP command on your system.
9. List the steps needed to save a drawing on your system.
10. Where is the location of the origin on your system?
11. What is the purpose of the MOVE command?
12. List a few examples of existing points that can be used for point selection.
13. Show how the last referenced point is represented on your CAD system.
14. List the steps used to format a disk on your system if it uses disk for storage.
15. List the steps necessary to plot a drawing.

Chapter Six DRAWING EXERCISES

1. Draw construction lines of random angles and spacing. Make a hard-copy print.
2. Fill the screen with horizontal and vertical construction lines spaced 1″ apart. Make a hard-copy print.

3. Fill the screen with lines of random lengths and angles similar to those shown in Figure 6-12. Make a hard-copy print.
4. Draw a 2" x 4" rectangle. Make a hard-copy print.
5. Using X-Y coordinates, draw the object shown in Figure 5-16 from question 7 in Chapter 5.
6. Using the SCALE command, change the scale of the completed drawing in question 5 to ¾ scale.
7. Make a 3-view drawing of the V-Block shown in Figure 6-35. Use the proper line type and weight.
8. Draw the borders shown in Figures 6-42 (Border A) and 6-43 (Border B).
9. Using Border A, create the drawing shown in Figure 6-44 using a .50 inch grid.
10. Using Border A, create the drawing shown in Figure 6-45 using a .50 inch grid.
11. From the exercises that follow, draw those chosen by your instructor.

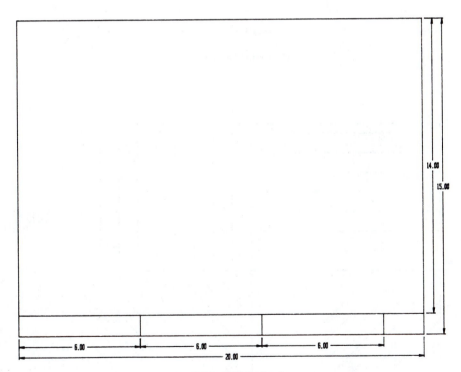

FIGURE 6-42 B size border.

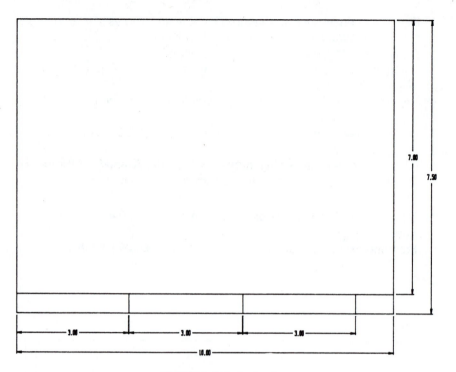

FIGURE 6-43 A size border.

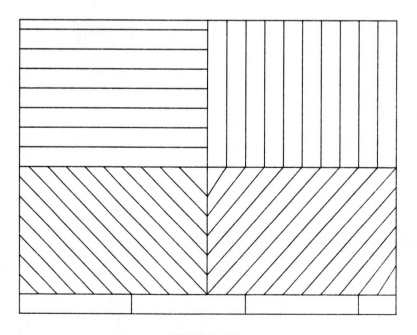

FIGURE 6-44 Lines.

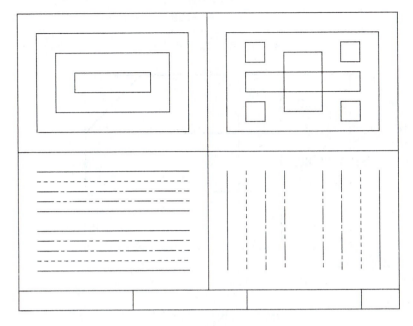

FIGURE 6-45 Lines and line styles.

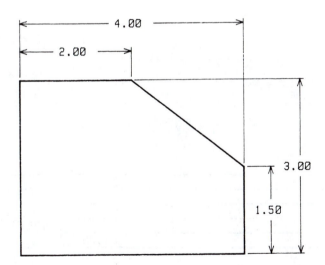

WEDGE BLOCK

FIGURE 6-46 Wedge block

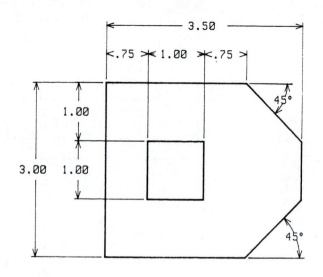

SUPPORT BLOCK

FIGURE 6-47 Support block

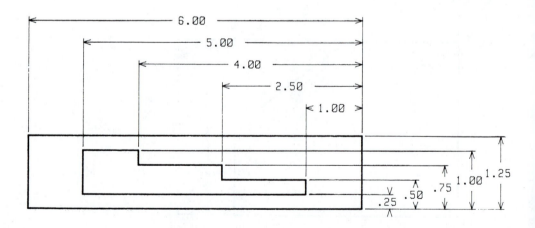

STEP CUTOUT

FIGURE 6-48 Step cutout

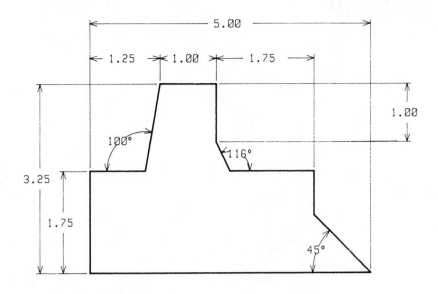

ANGLE PLATE

FIGURE 6-49 Angle plate

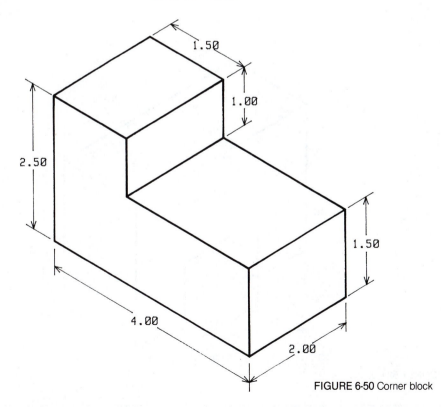

FIGURE 6-50 Corner block

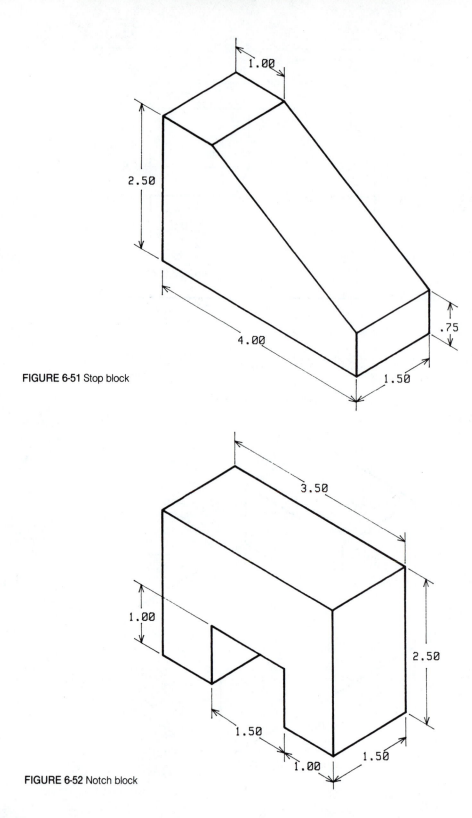

FIGURE 6-51 Stop block

FIGURE 6-52 Notch block

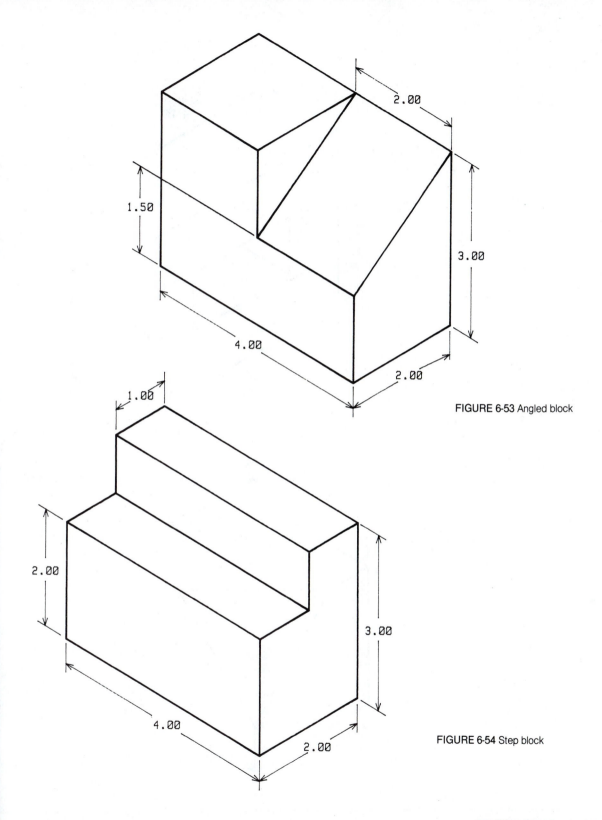

FIGURE 6-53 Angled block

FIGURE 6-54 Step block

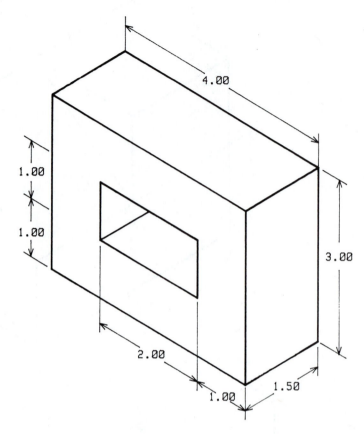

FIGURE 6-55 Rectangle

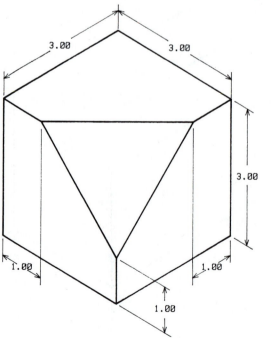

FIGURE 6-56 Cube

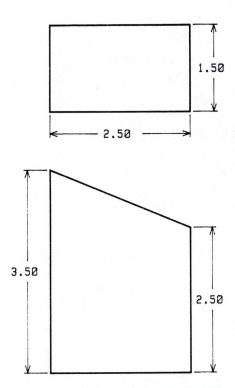

FIGURE 6-57 Develop this rectangular prism and add an auxiliary view of the oblique surface.

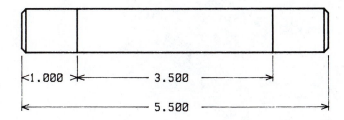

FIGURE 6-58 Construct schematic thread representations for the two ends of the shaft for unified national coarse, 10 threads per inch.

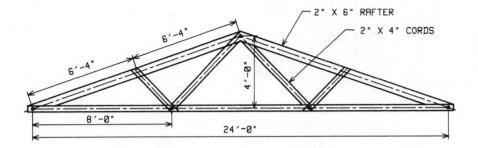

FIGURE 6-59 Draw this truss rafter scaling the drawing to fit on an 8-½'' x 11'' or 11'' x 17'' sheet.

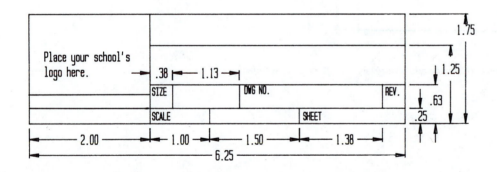

FIGURE 6-60 Title block.

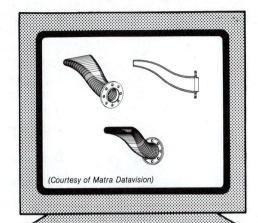

This chapter acquaints the user with additional special commands that are available on most CAD systems. These commands are useful for making CAD systems much more flexible in performing certain tasks shown in the earlier chapters. Two other functions introduced are symbols and parametric programming. You will learn how to create a symbol, and how symbols can be used to simplify dramatically the drawing of repetitive features. The editing of entities is also covered, showing how to use CAD to edit or change parts of a drawing without erasing and redrawing.

Special drawing features are also covered. The automatic drawing of slots, rectangles, and holes to user specifications is shown. The ability to mirror an object or part of a symmetrical object to complete a drawing is described. Another special drawing feature described is the duplicate or copy function. With this function, the user is able to copy part or all of a drawing, and make an exact duplicate of any scale to be placed anywhere on the drawing.

Another special function discussed is the "zoom" command. With this command, the user has the ability to enlarge a portion of a drawing to make it easier to draw detail work. You will also see how CAD can be used to move or "drag" an object or view to place it in a different position on the screen. The concept of levels is introduced, showing how to separate certain drawing details or entities.

These are only a few of the special features covered in this chapter. Some very powerful and time-saving functions are shown. Some of these features are so powerful and easy to use that users will find drawing to be enjoyable and their creative abilities freed because of the elimination of tedious drafting tasks. Some functions covered in this chapter will excite and challenge the user to try different methods and design ideas, because of the ease of making design changes with CAD. After completing

Chapter Seven
Editing, Symbols, Display Functions, and Files Handling

this chapter, you will have been exposed to the majority of functions available on most CAD systems. With the functions explained in the text and the drawings created by you, the prerequisite knowledge to use CAD and apply your newfound talents using this powerful tool will be yours. How to apply this tool and your ability with it is covered in Chapter Ten. But, first, you must master the special drawing functions described in this chapter.

PREREQUISITES

Before starting on this unit of instruction, you should be able to:

Complete

- a detailed drawing of a simple part.

Use

- the drawing functions presented in previous chapters.

OBJECTIVES

After completing this chapter, you will be able to

- draw special features such as slots, rectangles, holes, and ellipses.
- mirror an object.
- duplicate or copy an area.
- use special drawing functions such as zoom, drag, rotate, blank and move.
- erase by groups, levels, area, last entity, all entities or part of a single entity.
- assign levels to drawing features.
- create a symbol and add the symbol to a library.
- place a symbol on a drawing.
- write a simple parametric program.
- edit parameters assigned to entities.
- move, duplicate, copy, erase and archive files.
- list the general steps used to digitize a drawing.
- identify the various files management options available on CAD systems.

MENU ITEMS IN THIS CHAPTER

1. SLOT—command used to automatically draw a slot using parameters input by the user.
2. RECT—command used to draw rectangles by simply digitizing two corners.
3. HOLE—command used to draw circles with center lines.
4. ELLIPSE—command that will draw elliptical shapes of any specified major diameter and projection angle.
5. MIRROR—command used to produce a mirror image of a symmetri-

cal part around a given line.

6. COPY—command used to make a copy or copies of any defined area.

7. ROTATE—command used to rotate entities about an axis.

8. ZOOM—command used to enlarge or reduce the size of a drawing on screen without changing the dimensions in the geometric data.

9. UNZOOM—command which replaces the zoom view with the full view or unzoomed screen.

10. PAN—command used to move a drawing across the screen or up and down on the screen.

11. FULL SCREEN—command used to automatically fill the screen with the drawing.

12. MOVE—similar to the COPY command, except that the defined area is moved to a new location deleting the entities in the old location.

13. STRETCH—command used to drag or stretch an object by changing its length, width, or height.

14. GROUP—command used to combine entities into a defined group of entities.

15. SELECT LAYER—command used to change the active layer.

16. CHANGE LAYER—command used to change the layer assigned to an entity.

17. DISPLAY LAYER—command used to control the layers to be displayed on screen.

18. BLANK—command used to turn off specified screen images.

19. UNBLANK—command used to turn on specified images that were turned off using the BLANK command.

20. SAVE SYMBOL—command used to store a drawing as a symbol.

21. CALL SYMBOL—command used to place a stored symbol on a drawing.

22. SAVE PARAMETRIC—command used to store a parametric program.

23. EDIT—command used to change the parameters of specified entities.

24. DIGITIZE—command used to input an existing drawing into the CAD system.

25. FILES—command used to manage and manipulate drawings and data stored on the CAD system.

SPECIAL DRAWING FUNCTIONS

Many CAD systems have special drawing functions that allow the user to take shortcuts. These drawing functions are similar to using symbols but they do not have to be created by the user. They are provided to the user to make it easier to draw some of the more common features. Some of these special drawing functions are slots, rectangles, holes or circle center lines, parabolas, hyperbolas, and ellipses. Some CAD systems might have a command called POLYGONS. This command can be used to draw Polygons with up to 100 sides of equal length. Two very powerful functions that can be used with drawings created by the user are called "mirroring" and "duplicate" or "copy." Most CAD systems have many of the special functions described here. A careful study of the manual for your system will indicate which functions it has.

Slot

One special drawing function allows the user to create slots without drawing each entity making up the slot. The menu function used here to draw this feature is SLOT. When the user chooses this function, the computer prompts the operator to input the diameter of the arcs, the distance between the center points, and the angle at which the slot is to be drawn. The center position for the slot is then digitized to locate the slot on the drawing. Figure 7-1 and the following steps show how to use the SLOT function.

1. Select menu item SLOT.
2. The system will prompt you to input the width of the slot, the distance between the center points of each arc, and the angle at which the slot is to be drawn. For the example in the figure the width is .75", the distance between the center points is 1.50" and the angle is Ø degrees. Each parameter is input by using the RETURN command.
3. Digitize the desired position for the slot. In this example, the slot is placed by digitizing the center of the slot. The digitized point can be located over a grid point or construction point, or the X-Y coordinates can be input. The system will draw the slot to your specifications with center lines as shown.

You can see the advantages of using these special drawing functions for placing common drawing features. One of the biggest advantages to the drafter is the saving of time. The user simply selects the feature and inputs the parameters and digitizes the position, and the computer draws the feature.

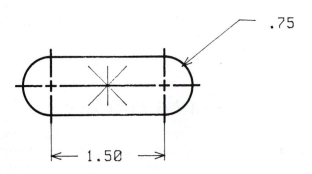

FIGURE 7-1 Drawing special entity "slot."

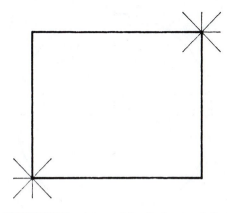

FIGURE 7-2 Drawing special entity "rectangle."

Rectangle

The rectangle is one of the most common features used in drawing. Because it is such a common feature, most CAD systems have an automated method of drawing it. This special function is used to draw rectangles to user specifications. Menu item RECT is used here for this function. After making this menu selection, the user digitizes opposite corners of the rectangle and the computer draws the rectangle using the digitized points as corners. For accuracy, the digitized points can be located over grid or construction points, or the user can input X-Y coordinates. Figure 7-2 and the following steps show how this function is used to create rectangles.

1. Select menu item RECT.
2. Digitize two corners. The computer will draw the rectangle through the two digitized points as shown in the figure. The size of the rectangle can be controlled by locating on grid points, construction points, referenced points, or X-Y coordinates.

Hole

When placing circles on a drawing they are usually used to represent holes drilled in the piece. Conventional drafting practice is for the placement of center lines through the center of the hole. By using the CIRCLE command, the user would have to change the line type and add center lines after the circle has been drawn. To eliminate the need to add center lines, some CAD systems give the user the option of drawing the center lines using a center line command. This command places the center lines over the center of the circle without having to change line types and locating each line separately. The user simply digitizes the circle and the center lines are added to the circle. Another method used to draw center lines for circles is a menu item that draws circles and center lines at the same time. This command is called HOLE, and its use is shown in Figure 7-3 and described as follows.

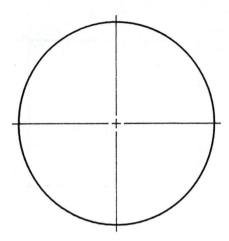

FIGURE 7-3 Placing center lines on holes.

1. Select menu item HOLE.
2. The computer will prompt for the desired diameter of the hole. Input the diameter followed by RETURN.
3. A prompt will ask the user to input the angle or desired rotation. This rotation angle is used to change the horizontal and vertical position of the center lines. In this example, zero is input followed by RETURN.
4. Digitize the location for the center of the hole. The computer draws the hole and places the center lines in one step as shown in the figure.

Ellipse

There are times when it is necessary to draw ellipses. This is especially true when making isometric drawings with drilled holes. Menu item ELLIPSE is used for this special function. This is a very time-saving command because of the number of steps it eliminates. The construction of ellipses using conventional drafting practices requires a number of steps or the use of a template, all of which can be eliminated using the ELLIPSE command. After choosing this command, the system prompts the user to input the major diameter of the ellipse. Another prompt will ask for the projection angle. The projection angle for an isometric drawing would be 35 degrees, for example. The angle of rotation then must be input. Finally, the location of the center point must be digitized. The system then locates the ellipse to your specifications at the digitized point. Figure 7-4 and the following steps show how the ELLIPSE command is used.

1. Select menu item ELLIPSE.
2. A prompt will ask for the major diameter. Key-in the diameter (1"), followed by RETURN.
3. A prompt will ask the user to input the projection angle. Key-in the angle (35), followed by RETURN.

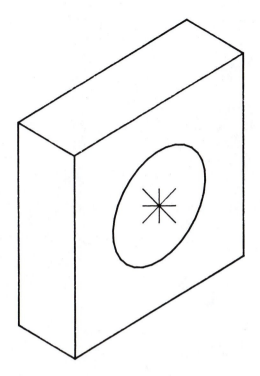

FIGURE 7-4 Drawing ellipses.

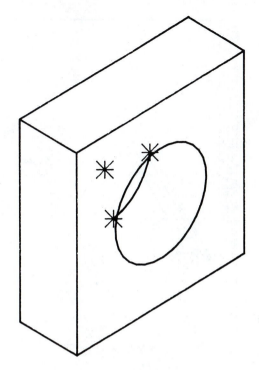

FIGURE 7-5 Drawing partial ellipses.

4. The system will now ask the user to input the angle of rotation. Key-in the angle (60), followed by RETURN.
5. The location of the center point is now digitized, followed by RETURN. The ellipse is drawn to your specifications. The RETURN command is used after digitizing the center to tell the computer that a full ellipse is needed.

If a partial ellipse is desired, the user could digitize the center point and the two end points for the partial ellipse, followed by RETURN. An ellipse such as the one in Figure 7-5, showing the back of the elliptical hole drawn in Figure 7-4, is an example of when a partial ellipse is used. The isometric exploded view shown in Figure 7-6 is an example of how ellipses are used in drawings to make an isometric exploded assembly.

MODIFYING GEOMETRY

Using the Mirror Command

The mirror command is a very powerful and time-saving function. It can be used for any object that is symmetrical. One half, one quarter or some fraction of the total object is drawn and the remaining parts of the

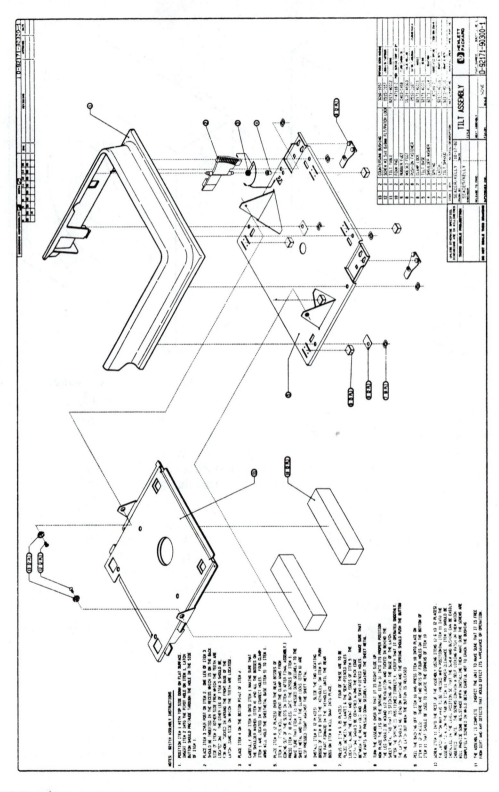

FIGURE 7-6 Isometric exploded assembly showing elliptical holes. (*Courtesy of Hewlett-Packard*)

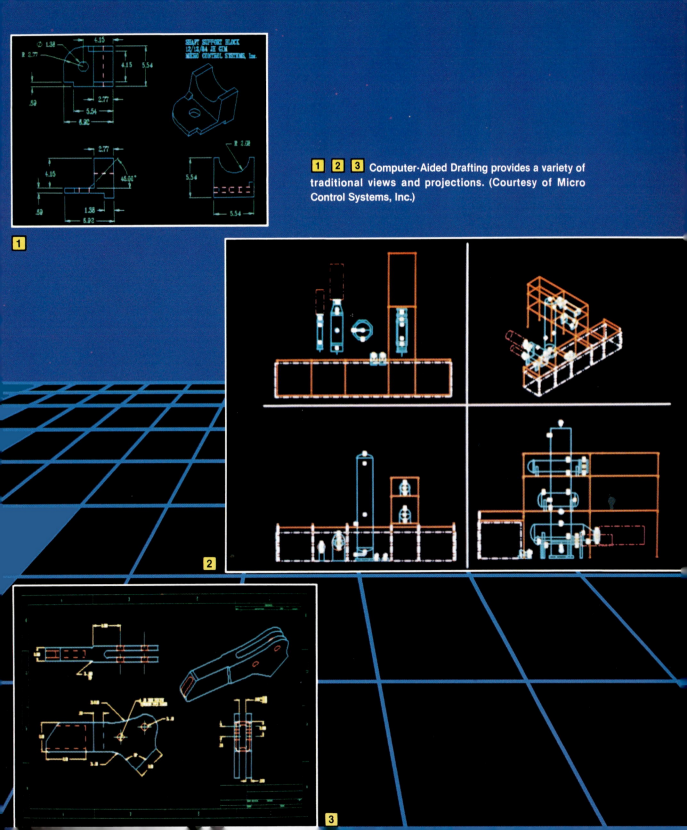

1 2 3 Computer-Aided Drafting provides a variety of traditional views and projections. (Courtesy of Micro Control Systems, Inc.)

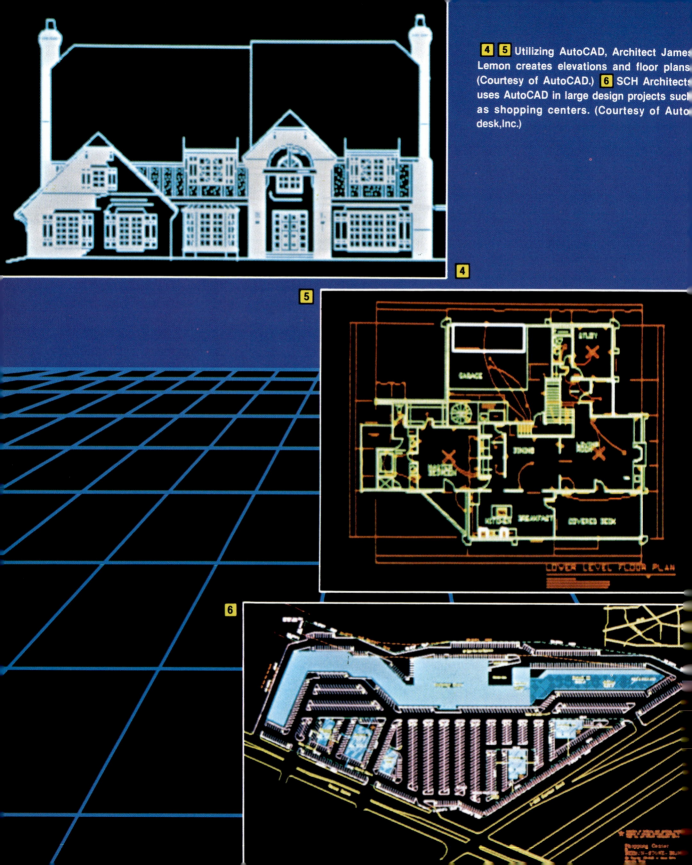

4 5 Utilizing AutoCAD, Architect James Lemon creates elevations and floor plans. (Courtesy of AutoCAD.) 6 SCH Architects uses AutoCAD in large design projects such as shopping centers. (Courtesy of Autodesk, Inc.)

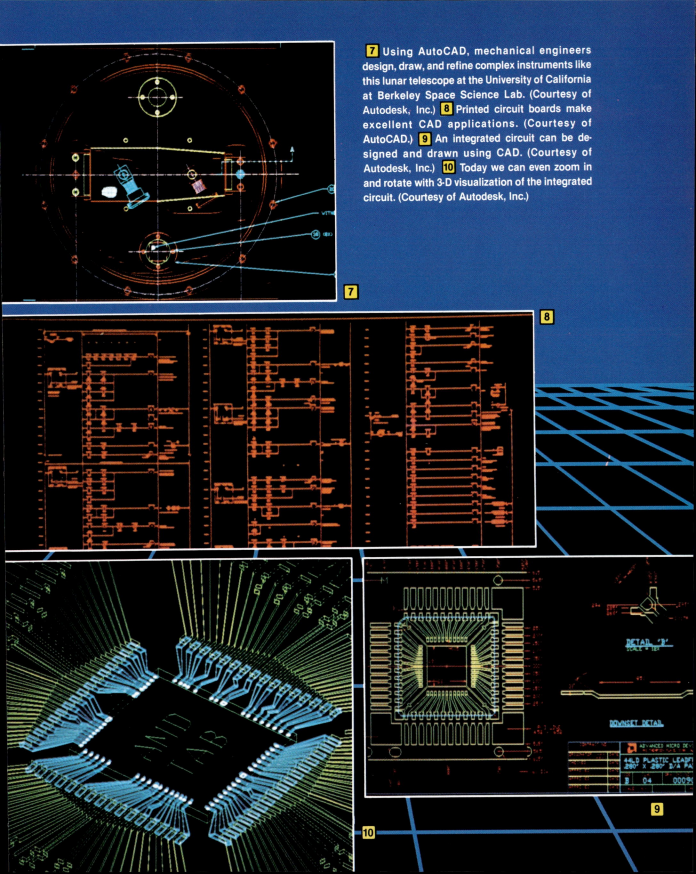

7 Using AutoCAD, mechanical engineers design, draw, and refine complex instruments like this lunar telescope at the University of California at Berkeley Space Science Lab. (Courtesy of Autodesk, Inc.) **8** Printed circuit boards make excellent CAD applications. (Courtesy of AutoCAD.) **9** An integrated circuit can be designed and drawn using CAD. (Courtesy of Autodesk, Inc.) **10** Today we can even zoom in and rotate with 3-D visualization of the integrated circuit. (Courtesy of Autodesk, Inc.)

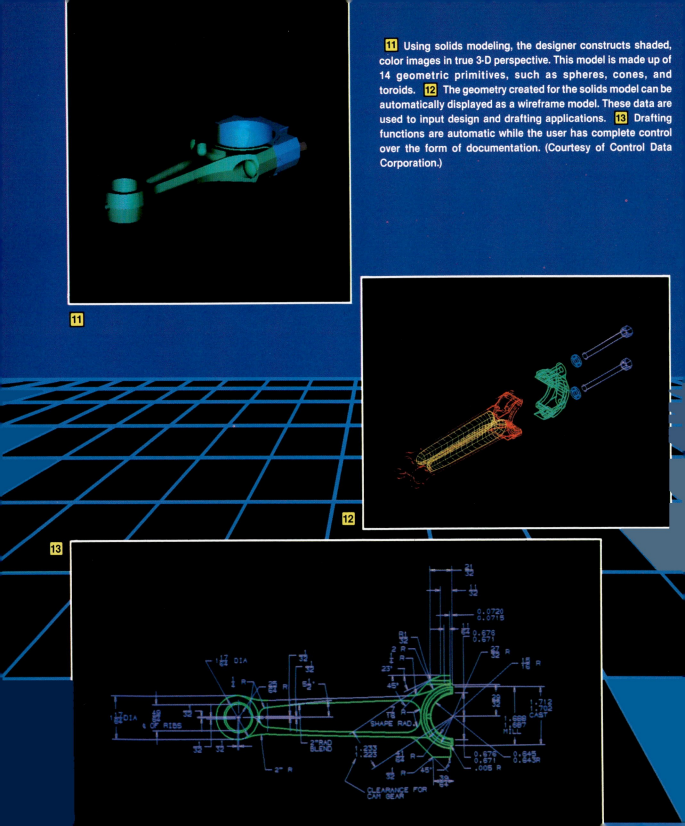

11 Using solids modeling, the designer constructs shaded, color images in true 3-D perspective. This model is made up of 14 geometric primitives, such as spheres, cones, and toroids. **12** The geometry created for the solids model can be automatically displayed as a wireframe model. These data are used to input design and drafting applications. **13** Drafting functions are automatic while the user has complete control over the form of documentation. (Courtesy of Control Data Corporation.)

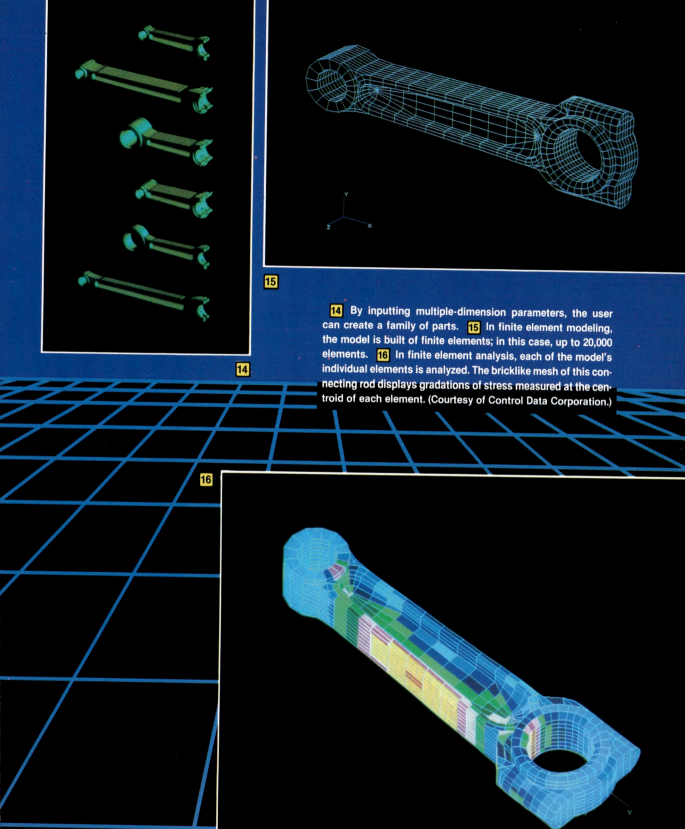

14 By inputting multiple-dimension parameters, the user can create a family of parts. **15** In finite element modeling, the model is built of finite elements; in this case, up to 20,000 elements. **16** In finite element analysis, each of the model's individual elements is analyzed. The bricklike mesh of this connecting rod displays gradations of stress measured at the centroid of each element. (Courtesy of Control Data Corporation.)

17 Design and production engineers can preview the cutter path for numerical control machining. This ensures the most accurate and precise paths before machining the actual part. **18** Numerical control (NC) capability allows generation of control tapes directly from design geometry. The NC output is used for the machining of the actual part. **19** The finished part. (Courtesy of Control Data Corporation.)

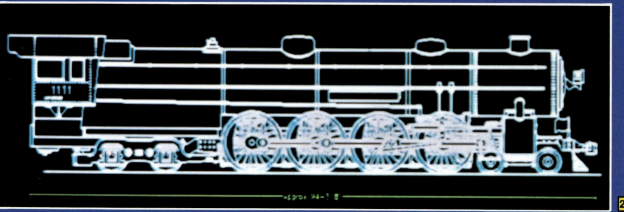

20

20 Large vehicles, or model locomotives like this one from Little Engines, Inc., are designed with AutoCAD. (Courtesy of Autodesk, Inc.) **21** Areas of a model locomotive are dynamically zoomed in for detailed design or plotting with AutoCAD. (Little Engines Inc.) (Courtesy of Autodesk, Inc.) **22** The three-dimensional drawing or "Solid Model" is functional as well as interesting. (Courtesy of Computervision Corp.)

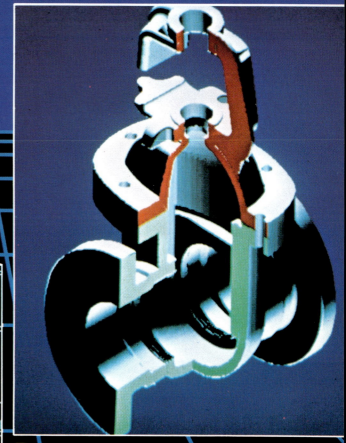

22

21

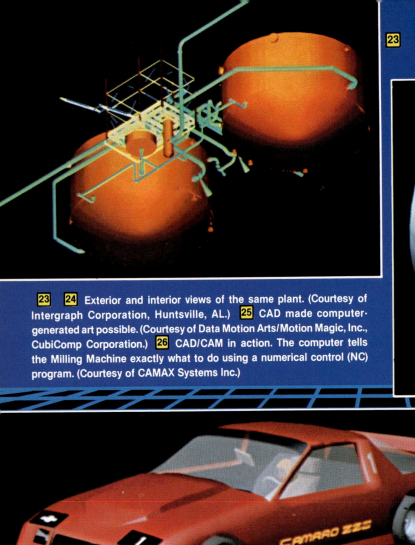

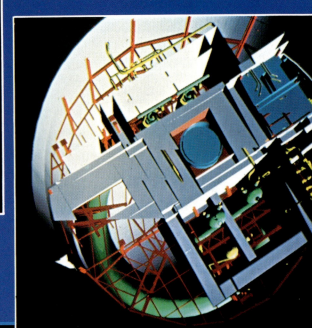

23 **24** Exterior and interior views of the same plant. (Courtesy of Intergraph Corporation, Huntsville, AL.) **25** CAD made computer-generated art possible. (Courtesy of Data Motion Arts/Motion Magic, Inc., CubiComp Corporation.) **26** CAD/CAM in action. The computer tells the Milling Machine exactly what to do using a numerical control (NC) program. (Courtesy of CAMAX Systems Inc.)

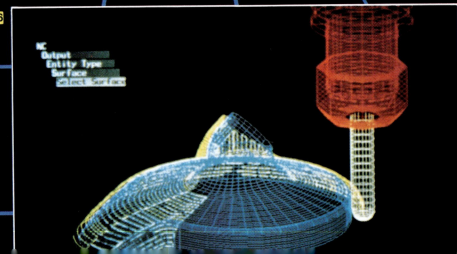

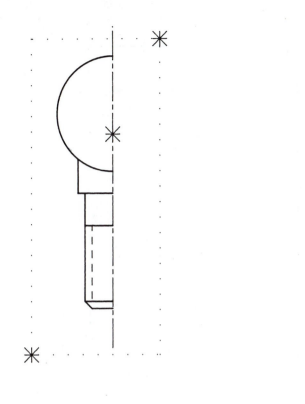
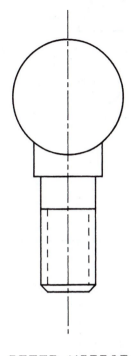

AFTER MIRROR
IS EXECUTED

FIGURE 7-7 Drawing half of a symmetrical part and using the MIRROR command to complete it.

drawing are mirrored to complete the view. This is accomplished by selecting menu item MIRROR and placing a box or fence around the drawing to be mirrored or selecting each entity to be mirrored singly. The fence is placed around the drawing by digitizing the two corners of a box which includes all the entities to be mirrored. The computer then draws a fence or box through the digitized points. The axis through which the object is to be mirrored is then digitized. The object is then mirrored about the digitized axis. Lettering may also be mirrored, but it will be drawn so that it is readable. Refer to Figure 7-7 and the steps that follow to mirror an object.

1. After part of the object is drawn, select menu item MIRROR.
2. Digitize two corners of a box that includes all the entities to be mirrored. A fence or box is drawn around the object.
3. Digitize the axis through which the object is to be mirrored. The object is mirrored, completing the view as shown in the figure.

There are times when only one quarter of the object has to be drawn. The mirror function can then be used to complete the object. Figure 7-8 and the following steps show how to mirror an object twice to complete a view having only one quarter of the drawing complete.

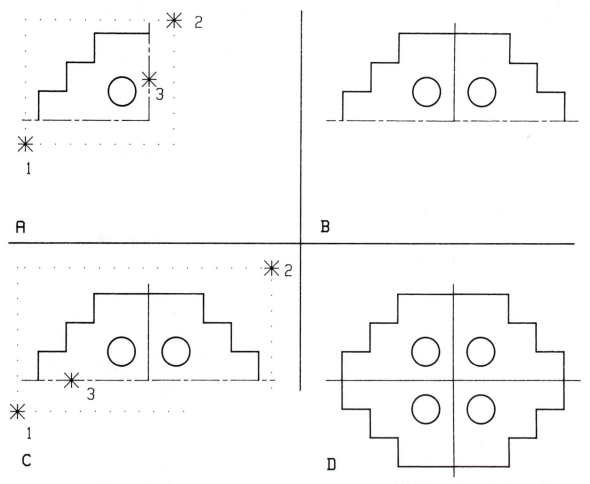

FIGURE 7-8 Drawing one quarter of a symmetrical part and using the MIRROR command to complete it.

1. After one quarter of the object has been drawn, select menu item MIRROR.
2. Digitize the corners of a box which includes all the entities to be mirrored. A box or fence is then drawn around the object by the computer as shown in part A of the figure.
3. Digitize the axis about which the object is to be mirrored. The computer mirrors the object as shown in part B of the figure. This completes one half of the view. Now the user can mirror the drawing shown in part B to complete the view.
4. Select menu item MIRROR.
5. Digitize the corners of the box which includes all the entities to be mirrored. A box or fence is then drawn around the view by the computer as shown in part C of the figure.
6. Digitize the axis about which the object is to be mirrored. The computer then mirrors the object as shown in part D, completing the view.

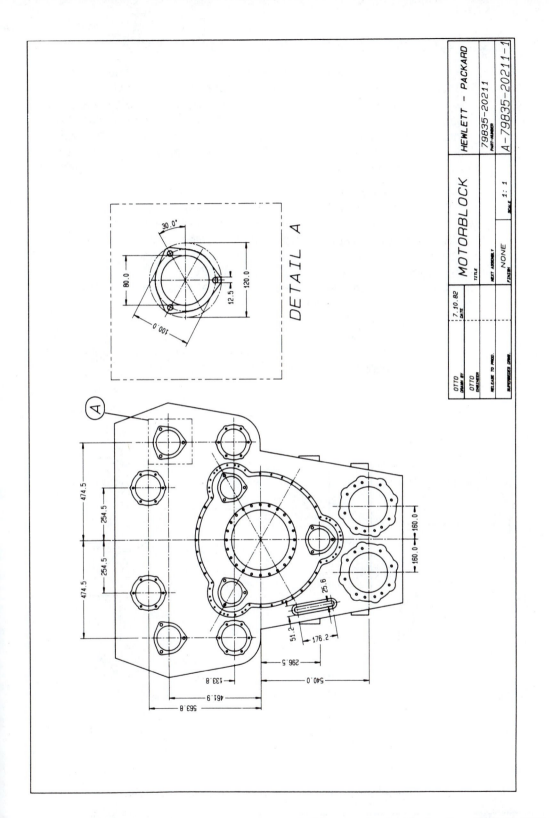

FIGURE 7-9 "Copying" and "scaling" a detail are easily accomplished with CAD. *(Courtesy of Hewlett-Packard)*

COPYING DETAILS IN A DEFINED AREA

The copy command is another very powerful command used with CAD. This command copies a specified view or part of a view to another area on the drawing. This time-saving command is used to copy and scale parts of views used to produce removed details, a common practice in many drafting applications, as shown in Figure 7-9. Menu item COPY is used to copy areas of a drawing. This function is shown in Figure 7-10 and explained in the following steps.

1. Select menu item COPY.
2. Digitize the corners of a box that includes all the entities to be copied. The computer then draws a fence around the area defined.
3. A prompt then appears asking for the scale of the desired copy. Example scales may be ⅟₁ for full, ⅗₁ for enlarging, and ½ to reduce the copy. Key-in the scale (¾₁) and RETURN.
4. A prompt then asks for the desired angle of rotation of the copy. Key-in the angle (Ø) followed by RETURN.
5. Digitize a point within the fence.
6. Digitize another point to locate the position for the point digitized in step 5 to be copied. The area within the fence is then copied to your specifications to the location digitized in this step.

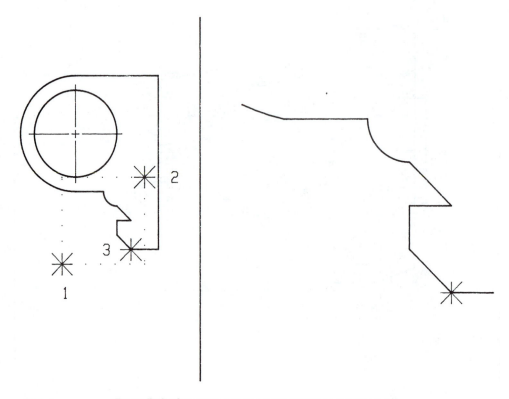

Figure 7-10 "Copying" and "scaling" details in a defined area.

MAKING MULTIPLE COPIES AND ROTATING

The Rotation Function

The ROTATION command is used to rotate selected entities about an axis. Entities are identified, the axis of rotation is digitized, and the angle of rotation is then input. ROTATE can be combined with other commands such as COPY to produce gears, bolt circles, and other parts. Figure 7-11 and the following steps demonstrate how ROTATE and COPY might be used to create a more complicated drawing.

1. Select menu item ROTATE.
2. Select those items to be rotated using WINDOW or digitizing each entity. For this example the square in the center of the part will be rotated by digitizing each entity.
3. A prompt will have you digitize the rotation origin, or "handle." Digitize a corner of the box.
4. A prompt will then ask you to input the angle of rotation. For this example, input 45 degrees. After the rotation angle is input, the selected entities will be rotated into their new positions as shown in Figure 7-11.

The duplicating function is used with the COPY command to make multiple copies of a view or a part of a view. The menu command used for this function is COPY. This function is selected after drawing the object to be duplicated. After COPY is selected, the number of copies is keyed-in. There is usually a limit to the number of copies that a system can make, although one to one hundred copies is not uncommon. Menu item COPY is then chosen and the area to be duplicated is digitized. The scale and angle of rotation is then input. The point that the first copy is to move is digitized. Copies, equally spaced, are then drawn according to the user's specifications.

Something quite remarkable occurs if the scale is changed when using the duplicate command. If a scale other than 1/1 is used, each succeeding copy is scaled from the previous copy. If a larger scale were chosen, each copy would grow; if a smaller scale were chosen each copy would shrink. The

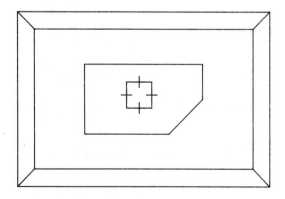 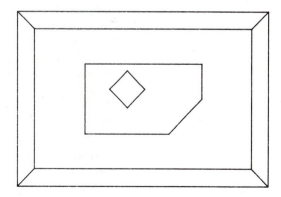

FIGURE 7-11 Using the rotate function to change the position of the square.

duplicate and copy function opens up a lot of different opportunities and applications for the user. These functions are not only time saving and useful both for the drafter and for design purposes, but they also allow the user to be much more creative such as in architectural renderings and other creative types of design work.

Figure 7-12 and the following steps show how to use the multiple copy command.

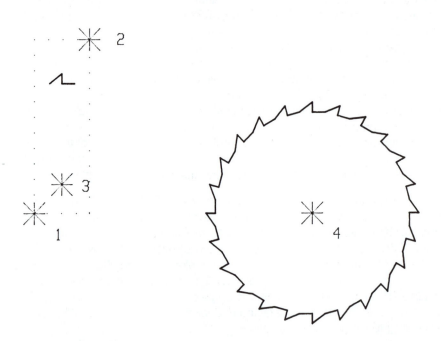

FIGURE 7-12 "Duplicating" and "rotating" a detail to produce a complete drawing.

1. Select menu item COPY.
2. A prompt will ask for the number of copies. Key-in the number (24) followed by RETURN.
3. Digitize the area to be duplicated by locating two corners of a box enclosing the area or singly digitize each entity. The computer then draws a fence around the area described.
4. A prompt will ask for the scale. Input the scale (1/1), followed by RETURN.
5. A prompt will ask for the rotation angle. Key-in the angle (15), followed by RETURN.
6. Digitize a point within the area to be moved.
7. Digitize the new location point for the copies. The computer then successively duplicates and rotates the object in the fence 24 times until the completed saw is drawn as shown in the figure.

Stretching a Drawing

The ability to stretch a drawing is a very handy function for use in design work. Stretching allows the operator to view different design alternatives quickly by changing the size of a design. Commands used for this function include "stretch," "move," or "drag," among others. In this example, STRETCH is used. To stretch a drawing, the operator must digitize two corners of a box that includes the entities to be stretched. The operator then digitizes a reference point in the box. The new location of the reference point is then digitized. The computer then moves the defined area to the new location and stretches the lines truncated by the defined area. *Truncated lines* are lines that are not totally within a defined area or fence. Figure 7-13 and the following steps show how the STRETCH command is used to stretch a drawing.

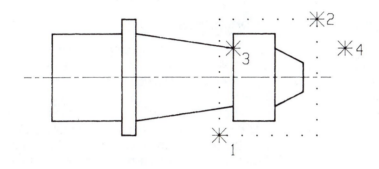

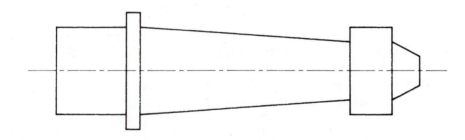

VIEW AFTER BEING STRETCHED

FIGURE 7-13 Using the STRETCH command

1. Select menu item STRETCH.
2. Key-in ¹/₁ for the scale.
3. Define the area to be stretched by digitizing two corners of a box that includes the area to be moved. The computer then draws a fence around the area defined.
4. Digitize a reference point within the fence. (3)
5. Digitize the new location point. (4) The defined area is moved, and the truncated lines are stretched.

SPECIAL DISPLAY FUNCTIONS

The following commands are very useful in helping the operator to draw with CAD as the part is being created. Some of these functions are additional erase commands that are useful for certain situations. Other functions add to the clarity of the drawing on screen by allowing the user to move and enlarge or reduce the size of the window or viewing area on screen. A *window* can be defined as the current viewing area on screen. Changing the window area changes the size or location of the drawing on the CRT. Changing the window does not change the size of the drawing as does SCALE. SCALE changes the size of the plotted drawing; changing the window temporarily changes the viewing area on screen. Another special display function allows the user to separate entities by views or levels.

The Zoom and Unzoom Commands

The zoom and unzoom commands are among the most useful functions that an operator can learn. With the zoom command, the user can enlarge or reduce the size of the drawing displayed on the screen without changing the geometric data base. This means that a drawing appears larger or smaller on screen, but the original scale is still retained in the computer's memory. This function is used to zoom in on small details that would be hard to see on screen. The ZOOM command is selected, and the user digitizes two corners of a box or fence. The area inside the fence is then enlarged and displayed on screen after the command is executed. After the view is enlarged, any detail work can be performed using the same scale as was used before zoom.

After the detail work has been completed, the unzoomed view can be put back on screen by using such commands as "unzoom," "fit view," or "full view." These commands will recall the size of the view before the ZOOM command was used. The command used in this text is called UNZOOM. When this command is selected, the screen is redrawn showing the view before zooming. Any detail work performed during the zoom view is included on the redraw.

Figure 7-14 and the following steps show how these two commands are used and the on screen results.

1. Select menu item ZOOM.
2. Digitize two corners of a fence for the desired area to be displayed on the screen. The corner is enlarged on screen allowing the user to perform such detail work as placing a rounded edge on the corner.
3. After the detail work has been completed, select menu item UNZOOM. The whole view is displayed on screen with any detail work completed in ZOOM carried over to the full screen view.

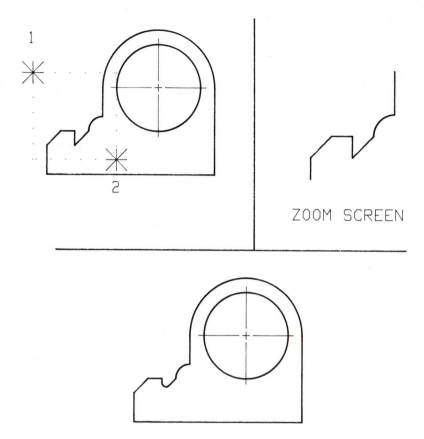

FIGURE 7-14 Using the ZOOM command to enlarge a detail on the screen. Any changes or additions to the drawing while in ZOOM will be displayed in the full view.

The Pan Function

The pan function is related to the zoom command in that it changes the drawing's position on the screen. However, the PAN command only moves the drawing up and down or left and right on the screen without changing the size of the drawing as does zoom. The pan function is executed by using special keys on the keyboard or through a menu command. Executing this command causes the current screen or window to change or pan to an area of the drawing. The amount of screen area panned is usually a default value that can be controlled by the operator. The amount of screen area panned or moved on CAD systems is normally about 50%.

The PAN command can be used while the view is enlarged or reduced with the ZOOM command. PAN is a very useful command in that it allows the user to move the drawing on the screen quickly without using the ZOOM command. Figure 7-15 shows an example of PAN being used to move the drawing on the screen to allow room for a top view.

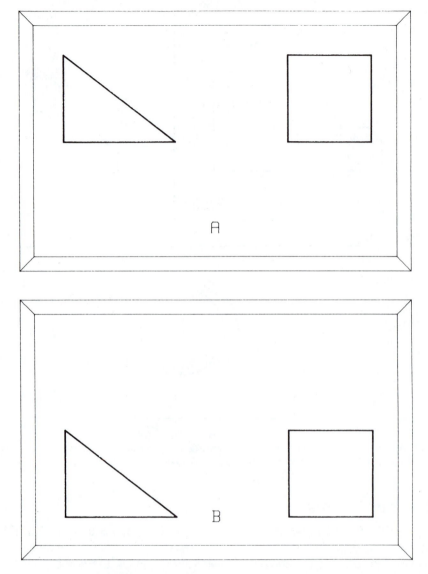

FIGURE 7-15 Screen B shows the position of the drawing in relation to its original position in screen A after the PAN function was executed.

Another method used to pan objects is to digitize a new center for the window or digitize two points. The difference between the two digitized points will be the distance the view will pan.

To relocate a drawing, the user digitizes a point on screen then digitizes another point on screen. The distance between the digitized points determines the distance that the viewing area moves on screen. Figure 7-16 and the following steps show how this PAN command is used to move the viewing area.

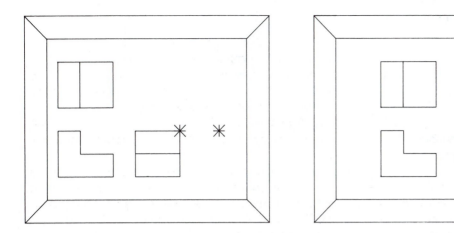

FIGURE 7-16 Using the TRANSLATE command to change the window or the location of a drawing on screen.

1. Select menu item TRANSLATE.
2. Digitize a point on the screen.
3. Digitize the new position for the first digitized point. The screen image will be deleted and replaced in the new position as shown in the figure.

The Full Screen Command

Another function that can be used to enlarge a drawing on screen is a full screen, auto-scale, or fill screen command. Executing this function causes the drawing on screen to fill the entire area on screen. As with the zoom command, the geometric data base is not changed. Dimensions remain the same, and any new entities added will be of the original scale of the drawing before full screen was executed. Figure 7-17 shows how the FULL SCREEN command is used. The user has only to make the menu selection and the computer automatically enlarges the drawing to fill the screen.

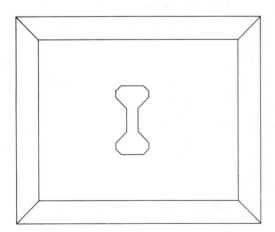 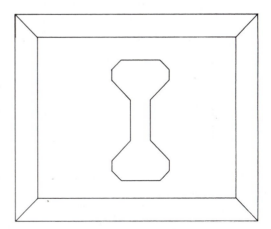

FIGURE 7-17 The FULL SCREEN command is used to enlarge the size of the drawing on screen.

Other Display Functions

Most CAD systems have a number of different ways of changing the window on a drawing. One of these other options is the WINDOW-HALF or ZOOM-HALF command. This command will automatically reduce by one half the displayed size of the part. This is demonstrated in Figure 7-18.

The ZOOM-DOUBLE or WINDOW-DOUBLE option will automatically double the display size of the drawing shown on screen. This is demonstrated in Figure 7-19.

Another option available with most CAD systems is the ZOOM-SCALE or WINDOW-SCALE command. This command will change the displayed size of the part to any scale input by the user. Entering a 2 will double the displayed size of the part; entering .25 will reduce the displayed size of the drawing to one fourth. Figure 7-20 demonstrates a drawing display which is reduced to one fourth its original size.

Many CAD systems will also have a method of recalling previous windows that have been used. Typically a system will be able to retrieve the last three or four windows that have been used on a drawing. This rolling stack of windows can be displayed one at a time on screen using a command such as WINDOW-BACK or ZOOM-BACK. Selecting this option once will cause the display to change to the previous window. For example, Figure 7-20 would change from quarter size back to full size by invoking this command.

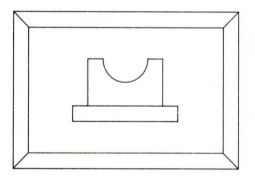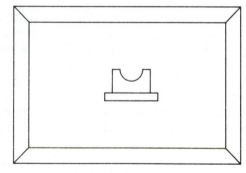

FIGURE 7-18 The ZOOM-HALF option is used to reduce the size of the displayed object by one half.

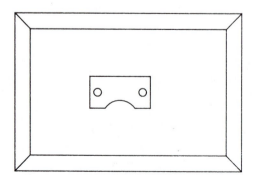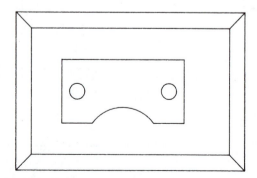

FIGURE 7-19 The ZOOM-DOUBLE option is used to double the size of the displayed object.

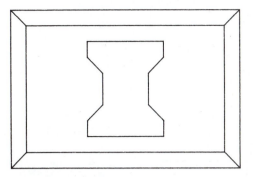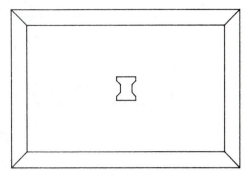

FIGURE 7-20 Using the ZOOM-SCALE option to reduce the size of the displayed object.

The Drag or Move Function

The MOVE or DRAG command is another very powerful command that can be used in a variety of ways to change a drawing. This function is very similar to the COPY command, except the defined entities are moved to a new location deleting the entities from the old location. On some systems this is a powerful design command that can change a drawing's dimensions by stretching the view. This command can also be used to move a view or part of a view to another position on the drawing. It can also be used to move text to a new position. MOVE is the command used here to implement this function.

Using the MOVE command, an area containing the entities to be moved is defined by digitizing two corners of a box or fence. Entities can also be identified singly, in a chain or by group. The user then has the option of scaling the area to be dragged. A reference point inside the defined area is then digitized. The user then digitizes the desired position for the new location of the reference point. The defined area is then moved from the original position to the new position on the screen. Figure 7-21 and the steps that follow show how the MOVE command is used.

1. Select menu item MOVE.
2. A prompt will ask for the scale of the area to be moved. Input the scale, such as ¹/₁ for full, ¹/₂ for half or ²/₁ for double size, and RETURN. In this example, ¹/₁ is used.
3. Define the area to be dragged by digitizing two corners of a box that includes the area to be moved (1,2). The computer then draws a fence around the area defined.
4. Digitize a reference point within the fence (3).
5. Digitize the new location point of the reference point (4). The area will be moved to the new location point as shown in the figure. In this example, the right-side view was moved to a new position on the drawing to allow room for dimensions.

Grouping Entities

A special menu option allows the user to combine entities into a group for erasing, blanking, editing, copying, plotting, moving or some other

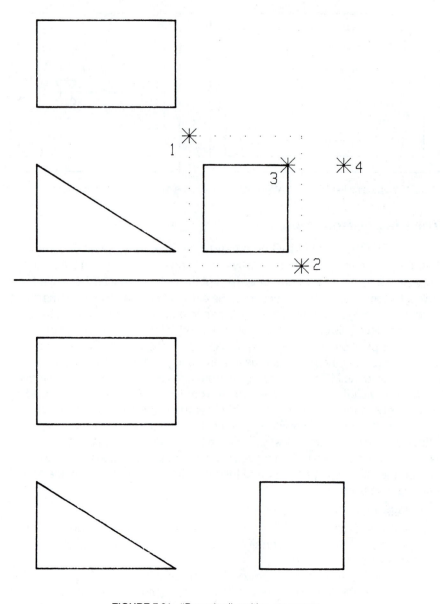

FIGURE 7-21 "Dragging" entities on screen.

function. With the GROUP command, the user has the ability to combine dissimilar entities into a common group. For example, the user could define all dimensions and fills as one group. Once a group has been defined, the user can manipulate these entities as a group instead of as separate entities. The entities will remain in a group until they are removed or ungrouped. The following steps and Figure 7-22 show how entities can be grouped.

1. Select menu item GROUP.
2. Select the entities to be grouped from the menu, such as LINES, TEXT,

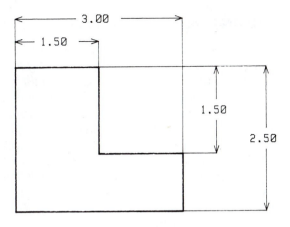

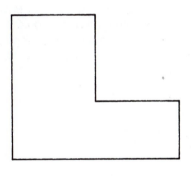

VIEW AFTER "ERASE"

"GROUP" IS EXECUTED

FIGURE 7-22 Grouping entities to erase or edit.

DIMEN, FILL or any other entity listed on the menu. In this example, menu items DIMEN and TEXT are selected. All dimensions and labels are considered to be a group.

3. The user can now manipulate the defined group. For this example, we are going to erase the group. This is done by selecting menu item ERASE.

4. Select menu item GROUP. By following the ERASE command with the GROUP function, all the dimensions and labels that were defined as a group are erased. This eliminates the need to digitize each entity or erase all dimensions, and then erase all labels.

Separating Entities by Levels or Layers

A common feature of CAD systems is the ability to segregate or group entities by "overlays," "levels," or "layers." A "layer" can be thought of as being similar to separate sheets of paper or overlays in conventional drafting terms. When drawing with instruments, it is common practice to divide a drawing into different parts by drawing them on separate sheets of paper. A level or layer in CAD can be thought of in the same manner. With CAD, the user is able to separate entities, views, line types, areas, and groups of entities by assigning them to different layers. See Figure 7-23. Most CAD systems have between 10 and 200 layers available to the user.

One layer is always active to the user when the system is turned on. That is, anything drawn will be assigned to the currently active layer. To change the active layer, menu item SELECT LAYER set, active, or some other option is used. The user makes the menu selection and keys-in the number of the layer to be active. The number of the current active layer is usually displayed at the bottom of the screen or in a prompt as a reminder to the user. Having entities assigned to different layers is a convenient method of separating entities. It allows the user flexibility in moving, changing, deleting, and plotting drawings.

Another command related to SELECT LAYER is a command to move or change entities from one layer to another. For example, suppose the user made a drawing complete with dimensions on the same layer but wanted to make two plots: one of the completed drawing and one without dimensions. With a menu item called CHANGE LAYER or MOVE, it is possible to

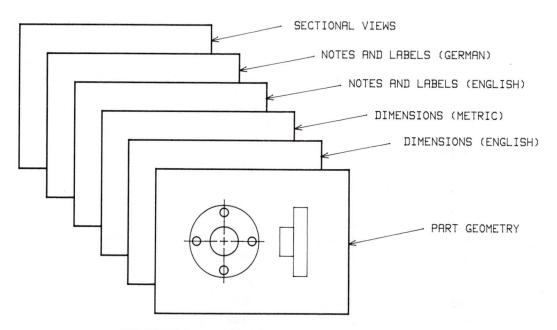

SECTIONAL VIEWS

NOTES AND LABELS (GERMAN)

NOTES AND LABELS (ENGLISH)

DIMENSIONS (METRIC)

DIMENSIONS (ENGLISH)

PART GEOMETRY

FIGURE 7-23 Example of how "layering" is used on a drawing.

group all dimensions and change the layer assigned to the dimensions. Most systems have the capability of making plots by levels so that the user is able to make two different plots: one with and one without dimensions if they were on different levels.

It is also possible to control which layers of a drawing will be shown on screen as well as which are plotted. Often, it is necessary to turn off certain entities on screen for clarity. If entities are separated by levels, the operator can turn off those entities not needed by using the VISIBLE or OFF command. With this command, the user keys-in the number of the layers to be seen on the screen. This does not change the active layer. Any entities added to the drawing will still be assigned to the active layer. The user also has the ability to erase a layer. Any entity assigned to a specific layer can be erased by using a menu item such as ERASE LAYER. This command allows the user to erase a layer by keying-in the number of the layer(s) to be erased.

Blanking Screen Images

When drawing with a CAD system, it is not uncommon to find the screen area filled with drawing entities, making drawing difficult and increasing redraw time. For this reason, systems have a BLANK, OFF, or HIDE command or some other method for temporarily turning off entities. This is not the same as deleting entities, because anything that has been blanked can be turned back on at any time. It should be noted that deleting an entity removes it from the CAD system memory and it is usually not retrievable. Blanking an entity simply turns off the entity until the drafter wants to see it again, similar to turning a room light on and off.

A number of different methods are used for turning off entities. Entities can be turned off individually, in groups, by line type, in a zone or by layers. Menu item BLANK is used to execute this command. To blank an entity the user chooses BLANK and digitizes the entity, as shown in Figure 7-24. To turn off an area, the user chooses BLANK and digitizes two corners of a box that includes the entities to be turned off, as shown in Figure 7-25.

Turning Entities On after Using the Blank Command

After turning entities off, the user must be able to turn the blanked items back on again. This is accomplished by using the UNBLANK command. To turn an entity on, the user selects UNBLANK and the item to be unblanked from the menu, such as CIRCLE. These steps cause the circle to reappear, as shown in Figure 7-26. Areas can also be turned back on by digitizing the corners of a box defining the area to be turned back on, as shown in Figure 7-27.

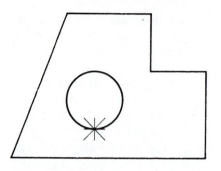

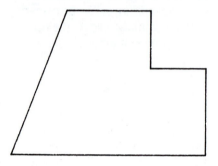

VIEW AFTER "BLANK" EXECUTED

FIGURE 7-24 Using the BLANK command to turn off entities temporarily

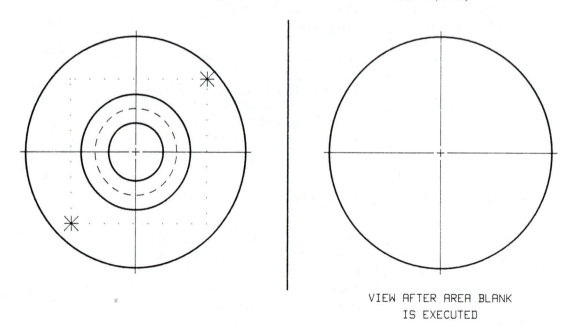

VIEW AFTER AREA BLANK
IS EXECUTED

FIGURE 7-25 "Blanking" entities in a defined area.

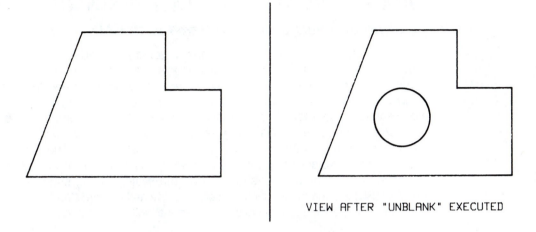

VIEW AFTER "UNBLANK" EXECUTED

FIGURE 7-26 Using the UNBLANK command to turn entities back on.

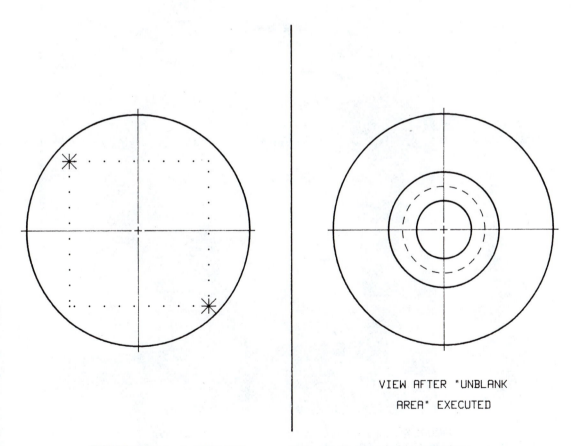

VIEW AFTER "UNBLANK
AREA" EXECUTED

FIGURE 7-27 Using the UNBLANK command to turn entities back on in a defined area.

CREATING, SAVING, AND PLACING SYMBOLS

Creating Symbols and Menu Items with CAD

In industry, many standard parts or items are used repeatedly on drawings. Working with traditional drafting tools, templates are used to draw common items. However, many industries require commonly drawn features for which no templates are made. With CAD, it is possible to draw an uncommon feature once, store it in memory or make it a menu item, and then retrieve it when needed. This is a great time saver when drawing standard parts, and it is one of the greatest advantages of using CAD. The function used for placing standard parts on a drawing is called "symbols," "cells," "macros," or "patterns," depending upon the system being used. In this text, SYMBOLS is the menu function used to create and retrieve standard items. A symbol is a graphic representation of a part that will be repeatedly used.

FIGURE 7-28 Symbols may be part of the menu, and are selected by digitizing the location on the tablet. *(Courtesy of Bausch & Lomb)*

When a CAD system is purchased, the user is provided with some symbols in the software. Other symbols can be purchased but, more commonly, symbols must be created by the user. If the CAD system has a menu on a tablet for item selection, the menu may have an area with a pictorial menu which can be digitized for symbol selection. Figure 7-28 shows a user digitizing an electronic symbol from the menu. Symbols not provided are easily made by the user.

To create a symbol, the user must draw the symbol and store it on the memory storage device. Once a symbol is stored, it can be retrieved by name and placed on the current drawing by digitizing the desired location. When a symbol is retrieved from memory, the user has the option of changing its size, rotating it, or mirroring it before placing it on the drawing.

The method used to create and retrieve symbols varies widely among CAD systems. Some systems do not make the distinction between symbols and any other type of drawing. With these systems, virtually any drawing produced can be retrieved and placed onto another drawing as if it were a symbol. On mainframe systems, there may be a file in memory used only for the storing of symbols. With this type of system, there may even be two types of files for the user: a private file and a public file. The *private file* is the file used by an individual user. The user can make additions and deletions from this file but other users do not have access to another's private file. A *public file* is one to which all users have access for retrieving symbols. However, some users may not be able to add symbols to this file, because this can only be done by certain privileged users having a necessary password.

Saving a Symbol in Memory

Because the methods used to make and retrieve symbols are so varied, the following steps are only a rough guide as to how symbols may be created with CAD.

1. After the symbol has been drawn, the user must store the drawing in memory. For some systems, the drawing is simply given a name and saved in the same manner as any other drawing, except that it may be put in a special file or disk used to store symbols. For other systems, the user must select a menu item, such as SAVE SYMBOL.
2. To save a symbol, the user must digitize a point or points on the drawing to be used as reference points or handles. The location of these points is important, because the symbol is retrieved and placed on a drawing by digitizing the location of the reference points. Figure 7-29 shows two reference points digitized on the one-cell battery symbol to be stored. The obvious reason for locating the reference points on the ends of the lines is that the battery symbol will be placed in alignment with a line representing a circuit. Without the proper reference points for a symbol, it would be very difficult to place a symbol in the proper position on a drawing. So the designer must think ahead and try to visualize the best location for placing reference points on the symbol.

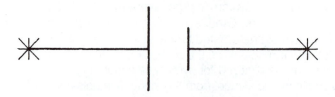

FIGURE 7-29 Storing a drawing as a symbol by digitizing reference points.

3. The user must then key-in a name for the symbol, followed by RETURN. In the following example, the symbol is called ONE CELL BAT.

Placing Symbols on Drawings

The placement of symbols varies a great deal from system to system. The following steps are only a rough guide to show how to place a symbol on a drawing. Refer to Figure 7-30 and the following steps to place the symbol created in Figure 7-29. For some standalone CAD systems, the first step is to place the symbols disk in the disk drive.

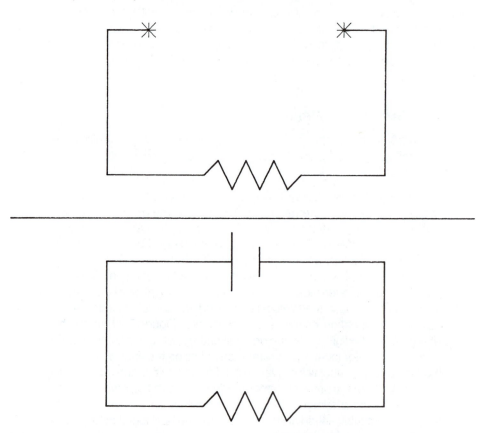

FIGURE 7-30 Placing a symbol on a drawing by digitizing the location of the symbol.

1. Select menu item CALL SYMBOL or RETRIEVE SYMBOL.
2. Key-in the title of the symbol, followed by RETURN. In this example, the user keys-in ONE CELL BAT.
3. The user now has the ability to scale, rotate, or mirror the symbol before placing the item by selecting the desired menu item.
4. The user must now digitize the desired location for the reference points or handles of the symbol. Remember that the reference points used for this symbol were located on the ends of the horizontal line. One end of the circuit line is digitized, followed by the second. After the second reference point is digitized, the symbol is drawn on screen, as shown in the figure. The user now has the option of accepting the position of the symbol where drawn, or changing the location.

PARAMETRIC PROGRAMMING LANGUAGES (MACROS)

Parametric programs or *macros* are English or programming languagelike statements chained together to perform a task. A *parametric program* assists the user in creating a part by prompting the user for input. Parametric programming languages (PPL) are an enhancement to standard graphics packages on some CAD systems. PPLs are used to simplify and automate certain drawing tasks and are an aid to nondedicated operators in using a CAD system.

Applications of PPLs include:
1. Parametric applications (family of parts).
2. To provide additional graphic functions that are not part of the software package, such as drawing a circle through three points.
3. Mathematical analysis of created geometry, such as mass properties calculations.
4. Creation of complex components, such as gears.
5. To automate repetitive and time-consuming tasks.
6. To standardize certain drawing procedures in an office.
7. To speed-up the creation and manipulation of part geometry.
8. To create additional text fonts.
9. To facilitate data processing and files management.

Most CAD systems give a name to their parametric programming language; for example, Computervision uses VARPRO, Gerber uses GOLD, CADKEY uses CADL, AutoCAD uses AutoLISP, VersaCAD uses CPL, Calma uses DAL, McAuto uses GRIP. Regardless of the name given, parametric programming involves the use of a language that uses graphic commands from the drawing software in addition to other programming commands. A parametric program used for drawing a part should prompt the user for the parameters of the part through inputs. Figure 7-31 lists a program written to create bolt circles. For example, suppose your CAD system does not have a menu item for drawing rectangles. If the CAD system had a PPL, the user could easily write a parametric program or macro to perform this task. The following is a program that might be written to create a rectangle of any size.

```
 10              ENTITY/C,P,CIR1,CIR2,CIR3,CIR4
 20    ASK:
 30              MASK/ALL
 40    TOP:      GPOS/'SELECT CENTER',X,Y,Z,RESP
 50              JUMP/TOP:,FINI:,,,,RESP
 60              PARAM/'BOLT CIRCLE DIA','DIA',BD,RESP
 70              JUMP/TOP:,FINI:,,RESP
 80              C=CIRCLE/X,Y,BD*.5
 90              PARAM/'HOLE LOCATION','ANGLE',A,RESP
100              JUMP/ASK:,FINI:,,RESP
110              MASK/ALL
120              NPTS=6
130              PARAM/'NUMBER OF INTERVALS','NPTS',INT,NPTS,RESP
140              JUMP/ASK:,FINI:,,RESP
150    QU:       CHOOSE/'TYPE OF HOLE','DRILL THRU','TAP',RESP
160              JUMP/ASK:,FINI:,,,,TAP:,RESP
170    RDA:      PARAM/'REQD DIA.','HOLE DIA.',D,'CBORE DIA.',CD,RESP
180              JUMP/QU:,FINI:,,RESP
190              CHOOSE/'TYPE OF FONT','SOLID','DASHED',RESP
200              JUMP/RDA:,FINI:,,,,DAS:,RESP
210              FONT/SOLID
220              JUMP/DRC:
230    DAS:      FONT/DASH
240    DRC:      ANG=360/NPTS
250              IF/NPTS-2,ASK:,,
260              IF/CD,RDA:,LOOP2:,
270              DO/LOOP1:,I,1,NPTS
280                GANG=A+I*ANG
290                P=POINT/C,ATANGL,GANG
300                CIR1=CIRCLE/CENTER,P,RADIUS,D*.5
310                CIR2=CIRCLE/CENTER,P,RADIUS,CD*.5
320    LOOP1:
330              JUMP/STOP:
340    LOOP2:    DO/LOOP3:,I,1,NPTS
350                GANG=A+I*ANG
360                P=POINT/C,ATANGL,GANG
370                CIR1=CIRCLE/CENTER,P,RADIUS,D*.5
380    LOOP3:
390              JUMP/STOP:
400    TAP:      CHOOSE/'TYPE OF FONT','SOLID','DASHED',RESP
410              JUMP/QU:,FINI:,,,,R2:,RESP
420              DAST=1
430              JUMP/R1
440    R2:       DAST=2
450    R1:       CHOOSE/'SELECT ONE','STD #','STD DECIMAL',RESP
460              JUMP/TAP:,FINI:,,,,RPL:,RESP
470              CALL/'STD#1,SIZE,TPI,REJ
480              IF/REJ-1,,R1:,
490              L8=1/TPI*,61343
500              JUMP/DTP:,DAH:,DAST
510    RPL:      CALL/'STDDECIMAL1',SIZE,TPI,REJ
520              IF/REJ-1,,R1:,
530              L8=1/TPI*.61343
540              JUMP/ ,DAH:,DAST
550    DTP:      ANG=360/NPTS
560              IF/NPTS-2,ASK:,,
570              DO/LOOP4:,I,1,NPTS
580                GANG=A+I*ANG
590                P=POINT/C,ATANGL,GANG
600                FONT/DASH
610                CIR3=CIRCLE/CENTER,P,RADIUS,(SIZE/2)
620                FONT/SOLID
630                CIR4=CIRCLE/CENTER,P,RADIUS,(SIZE/2)-L8
640    LOOP4:
650              JUMP/STOP
660    DAH:      FONT/DASH
670              ANG=360/NPTS
680              IF/NPTS-2,ASK:,,
690              DO/LOOP5:,I,1,NPTS
700                GANG=A+I*ANG
710                P=POINT/C,ATANGL,GANG
720                CIR3=CIRCLE/CENTER,P,RADIUS,(SIZE/2)
730                CIR4=CIRCLE/CENTER,P,RADIUS,(SIZE/2)-L8
740    LOOP5:
750    STOP:     DELETE/C
760              JUMP/ASK:
770    FINI:
780              HALT
```

FIGURE 7-31 Macro program printout used to design bolt circles and the resulting drawing. *(Courtesy of STAMCO)*

1. Start
2. PROMPT. Enter the width and the height of the rectangle.
3. INPUT = W,H
4. LINE
5. X = ∅ Y = ∅
6. X = W Y = ∅
7. X = ∅ Y = H
8. X = −W Y = ∅
9. X = ∅ Y = −H
10. END

This program would draw a rectangle to the input dimensions. The program begins with a statement to start the program running. Step 2 is the statement used to have a prompt appear on screen so that the user can input the necessary dimensions. Line 3 sets the input values equal to W for the width and H for the height. Step 4 calls on a command from the CAD software package, menu item LINE, to draw the lines for the rectangle using the values input in step 2. Line 5 sets the digitized point for locating the rectangle on the drawing as the origin. Step 6 draws a horizontal line using X-Y coordinates, X being equal to the input value of W, and Y being equal to ∅. Step 7 will draw a vertical line from the end of the horizontal line. Step 8 will draw a horizontal line to the left from the end of the vertical line, and step 9 will draw a vertical line back to the origin. The END statement tells the computer to end the program.

Once a program has been written, it can be added to the menu as another menu option or it can be added to the symbols library or a separate file for parametric programs using a command such as SAVE PARAMETRIC. If it is added to the menu, then the user has only to digitize the menu location and the program will start running. If it is added to a library on file, the user must activate the program by entering the name of the program. The user will then digitize the location of the origin and the program will start running beginning with the prompt statement.

The example described is a very simple task solved with a parametric program. Much more complicated programs may be written by the user as experience is gained. More complicated drawings can include the drawing of orthographic views of the object, dimensioning and adding notes to the object, and placing a title block on the drawing with the appropriate text. These can all be done by simply answering a series of prompts from a parametric program. The use of PPL is an important addition to any CAD system. As a CAD operator, you will find that learning to write parametric programs will be easier if you have had a programming course. However, most PPLs are easy to learn, and usually require little programming experience to write simple programs. Figure 7-32 shows three different examples of macros.

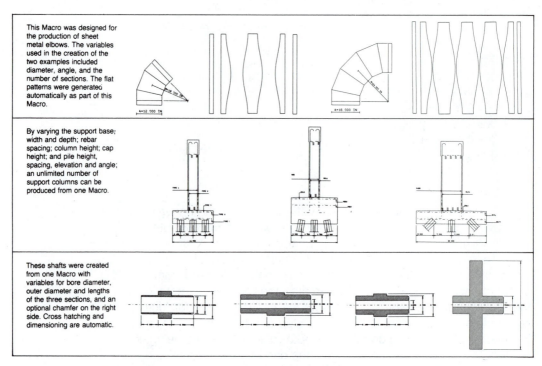

This Macro was designed for the production of sheet metal elbows. The variables used in the creation of the two examples included diameter, angle, and the number of sections. The flat patterns were generated automatically as part of this Macro.

By varying the support base; width and depth; rebar spacing; column height; cap height; and pile height, spacing, elevation and angle; an unlimited number of support columns can be produced from one Macro.

These shafts were created from one Macro with variables for bore diameter, outer diameter and lengths of the three sections, and an optional chamfer on the right side. Cross hatching and dimensioning are automatic.

FIGURE 7-32   Macros used for sheet metal elbows, support bases, and shafts. *(Courtesy of Holquin)*

AUTOMATED DATA RETRIEVAL

Many CAD systems are able to generate automatically a bill of materials and other data sheets from related drawings. This special feature extracts information from drawings on file and combines it into a bill of materials or data tables. Information about each part is input along with the part or subassembly name. Typical information in a bill of materials includes part number, quantity, description, size, material, vendor, cost, and notes. This information is gathered by the computer for an individual drawing or a group of subassembly drawings.

The computer searches the drawing files looking for part numbers and matches them against the part or subassembly name. This information is compiled and put in a table listing the information for each part. If a drawing is changed, the tables will automatically be changed. This means that the table is automatically updated as the drawing file is changed. For example, in designing a house, a client may want to determine what the difference in cost would be if the number of windows on the house were to be changed. The table is automatically updated as the changes are made on the drawing, including the cost. The table can then be output to the printer or plotter for a hard copy. Figure 7-33 shows tables that were created from the graphic data. The information shown in these tables can be used to determine project costs, vendor costs, where parts are used, family of parts, cost feasibility of

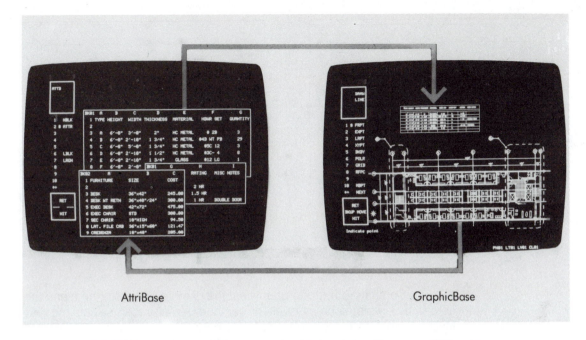

AttriBase   GraphicBase

FIGURE 7-33 Data retrieval from the graphics data base is possible on many CAD systems. *(Courtesy of Bruning CAD)*

design changes, project scheduling, process planning, inventory control, budgets, material requirements planning (MRP), and more.

EDITING A DRAWING

Many times while making a drawing the user will find that changes must be made because of design changes or simple operator error. If the changes are minor, using the ERASE command may be the quickest method to make changes. However, if a number of changes must be made, using the system's editing capabilities may be easier and faster. Editing can be used to change line types and thicknesses, edit text, change pen numbers, change entities to different levels, or to change virtually any parameter associated with an entity. Menu command EDIT is now added to the list of menu items in this text.

To edit an entity, the user chooses menu item EDIT or MODIFY, then selects the menu item to edit, followed by digitizing the entity to be edited. The user has the choice of editing all entities in a group or one at a time. Figure 7-34 and the following steps show how the editing command is used to change a solid line into a hidden line.

1. Select menu item EDIT.
2. Select menu item LINE TYPE.
3. Select menu item 2 for a hidden line.
4. Digitize the line to be changed. The computer then deletes the solid line and replaces it with a dashed line.

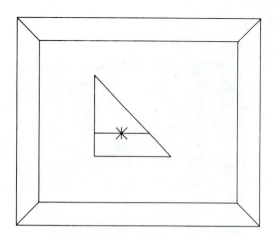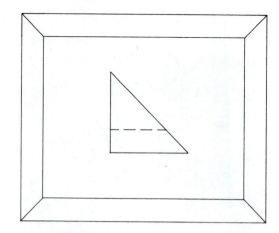

FIGURE 7-34 Drawing entities can be edited to change line type.

To change a group of entities, the user selects from the menu the item to be edited, such as levels or pen number, and then selects the new level or pen. The entities to be edited are then grouped using the GROUP command. All items in the group then change to the new parameters.

The editing function is done instead of erasing and redrawing entities, and it can save time when making changes to a drawing. It is a powerful function that varies widely from system to system. But, before changing a drawing, you should determine if it is best to erase and redraw or simply to use the edit function.

DIGITIZING A DRAWING

Many CAD systems allow the user the option of digitizing existing drawings or sketches instead of redrawing them completely with the system. By using the DIGITIZE function, you can input an existing drawing into the CAD system. To do this there must be a digitizer board and an input device to locate or digitize the position of various drawing entities. See Figure 7-35.

The digitizing function is usually accomplished by taping the drawing to the board, and moving a cross-hairs puck or stylus over the position to be input. The user then presses a "master button" and that point is input into the system. The drawing entity is then displayed on screen. The puck usually has a number of buttons for making menu selections without having to look at the screen or the menu is displayed on the digitizer. For example, button number 1 might be used to draw solid lines.

Figure 7-36 and the following steps show how to digitize a simple drawing.

1. Tape the drawing to the digitizer board.
2. Select menu item DIGITIZE.

FIGURE 7-35 Large digitizer board and multibutton puck used to input drawings into a CAD system. *(Courtesy of Summagraphics Corp.)*

3. Locate the drawing on the digitizer board by digitizing the two corners of the drawing sheet.
4. Digitize one corner of the drawing.
5. Select menu item LINE by pressing the proper key on the puck or selecting that command from the menu.
6. Digitize the end point of the line on the drawing. Line 3-4 will be drawn on screen.
7. Digitize the end point of the next line. Line 4-5 will be drawn on screen.
8. Digitize the end point of the next line. Line 5-6 will be drawn on screen.
9. Digitize the first point digitized. Line 6-3 will be drawn on screen completing the drawing to be digitized.

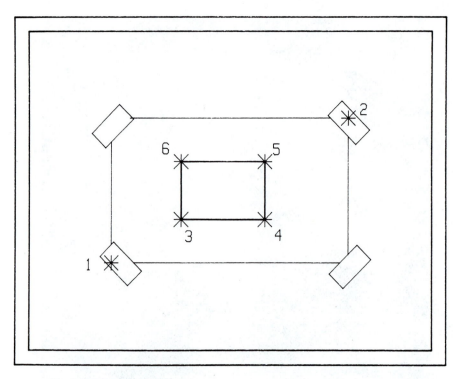

FIGURE 7-36 Digitizing a drawing by locating end points of the lines.

FILES MANAGEMENT

As the number of drawings saved on files grows, the operator will find it necessary to manipulate them for one reason or another. The files management function allows the user to manipulate files created on the CAD system. A CAD system has to handle thousands of files over a period of time, and must manipulate and catalog files that may be saved for years. File management functions include:

1. Creating drawing files (SAVE).
2. Resaving or updating a file.
3. Creating plotfiles (PLOT).
4. Expanding or contracting existing files.
5. Creating user library of symbols.
6. Recovering or rebuilding files that may be unreadable due to a power failure or an error.
7. Moving files from one library to another.
8. Protecting files so that access is limited to privileged users having a password.
9. Grouping files by any item information associated with the file (name, date, project number, and so forth).
10. Archiving drawings to offline backup.

11. Reporting a listing of the files in a library.
12. Keeping a log of users, updates, time signed in and out, and so forth.
13. Deleting a file from memory.
14. Copying a file.
15. Renaming a file.
16. Initializing or formatting a disk.

All of the foregoing functions are possible through files management. As can be seen, a number of different files management functions can be performed by the operator. Once a file is created, it can be moved, deleted, merged, grouped, copied, rebuilt, and archived. These are very important functions because you must be able to manage your drawings so that they are not lost or take a lot of time to retrieve. How to access these functions varies but most are accessed by a menu command. Menu command FILES is used here to show how to perform a files management function.

Files Management on Microcomputers

For some standalone microcomputers, floppy disk or microdisk drives are the only type of memory device used. For these systems, files management is relatively easy, because usually there is only one storage device to contend with. To archive or backup a disk, the user has only to copy the disk onto another. The use of floppy disks is a relatively safe method of storing files. Of course, files manipulation is slow when compared to other memory devices. For this reason, and others, some CAD systems use hard disk drives for memory storage.

For those systems having a hard disk drive, such as a Winchester drive, files management becomes a little more complicated. With a hard disk drive, it is necessary to have floppy disk drives for backup. The reason for this is that if the hard disk drive were to have a "fatal crash" most or all files stored on it would be lost. For this reason it is necessary to backup files using floppy disks or microdisks.

Files Management for Mini and Mainframe Computers

Most larger multistation CAD systems use disk packs for memory storage. These disk packs must be backed-up just as the hard disk drives should be. Backup is usually accomplished by archiving the files to magnetic tape. The magnetic tape is then stored in a safe environment for later use.

Because storing files on magnetic tape can be a tedious task that should be performed routinely every day, the drafter may not become involved. A systems manager trained in using computer hardware would probably be assigned to this task.

SUMMARY

Learning to use some of the special drawing features described and illustrated in this chapter will add greatly to your speed and flexibility. You should also be able to begin creating your own symbols file for any special drawing features that you expect to use more than once. In addition, you can start to clean up your files and maintain them properly for safety and speed of retrieval. You should now have all the important prerequisite skills necessary to create virtually any type of drawing. The skills you have learned will go a long way toward your becoming an experienced CAD operator capable of transferring your skills to other systems.

Chapter Seven REVIEW

1. Retrieve an old drawing and group some of the entities on that drawing.
2. After completing number 1, erase those entities as a group.
3. Practice drawing some special features used on your system, such as rectangles or center lines for circles.
4. Find a simple symmetrical object in a standard drafting textbook. Draw half of the object, and mirror it to complete the drawing.
5. Draw a 4" x 6" rectangle on screen. Copy that rectangle 10 times using a scale of .75.
6. Create a symbol and add it to your CAD system to start a symbols library.
7. Try writing a simple parametric program to draw the profile of a hex head bolt.
8. If you have a digitizer, digitize a simple drawing.
9. List some of the special drawing functions that your CAD system can create.
10. List the steps necessary to draw an ellipse on your CAD system.
11. List the steps used to copy and duplicate an object on your CAD system.
12. What are the names of the commands used on your CAD system to zoom and unzoom screen images?
13. Describe how the pan function works on your CAD system.
14. List the steps used to translate a drawing on screen.
15. List the steps used to drag an image on screen.
16. List the steps used to group entities on your CAD system.
17. List the steps used to create a symbol on your CAD system.
18. List the steps used to place a symbol on a drawing using your CAD system.
19. What are the names of the commands used to blank and unblank entities on your CAD system?
20. What is the name, if any, given to the parametric programming language on your CAD system?
21. List some of the files management functions on your CAD system.

Chapter Seven DRAWING EXERCISES

1. Create the symbols selected by your instructor.
2. Create a symbol for the title block produced in Figure 6-60, Chapter 6.
3. From the exercises that follow, draw those selected by your instructor.

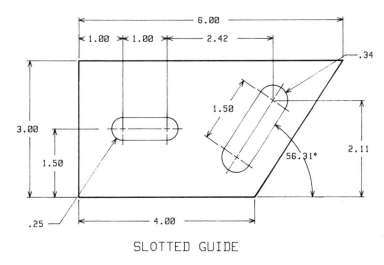

SLOTTED GUIDE

FIGURE 7-37 Draw the Slotted Guide and dimension.

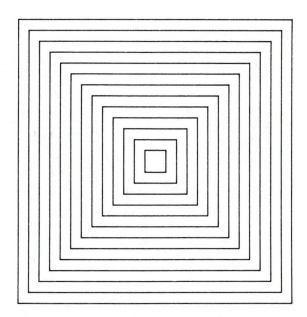

FIGURE 7-38 Use the COPY command to draw the squares which are spaced ¼ " apart.

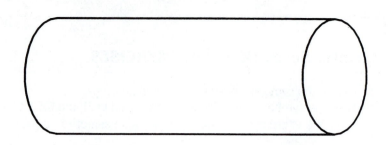

1" MAJOR DIAMETER. 4.25" LONG

SHAFT

FIGURE 7-39 Draw the Shaft using ellipses for the ends.

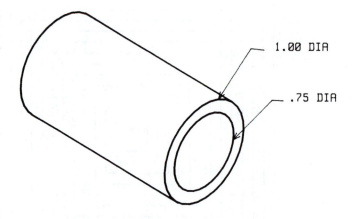

1.00 DIA

.75 DIA

2.5" BUSHING

FIGURE 7-40 Draw the 2.5" Bushing using ellipses.

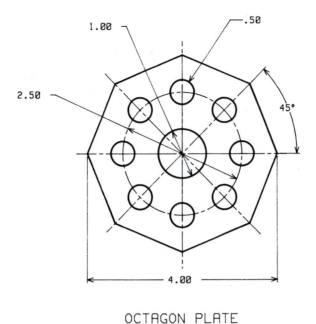

OCTAGON PLATE

FIGURE 7-41 Draw the Octagon Plate using COPY, MIRROR, and/or DUPLICATE commands.

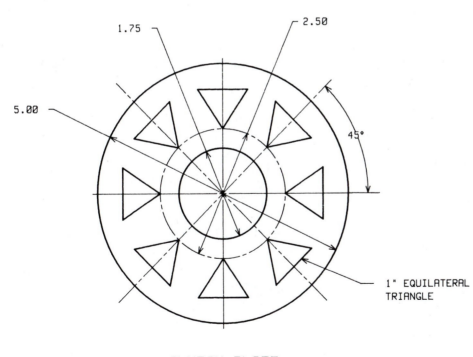

CLUTCH PLATE

FIGURE 7-42 Draw the Clutch Plate using COPY, MIRROR, and/or DUPLICATE commands.

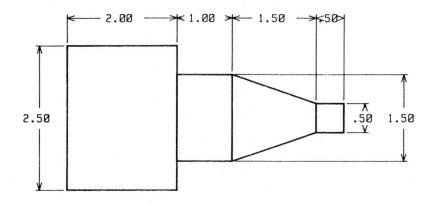

TAPERED ROD

FIGURE 7-43 Draw the Tapered Rod then stretch the 1.50″ angled portion to 3″.

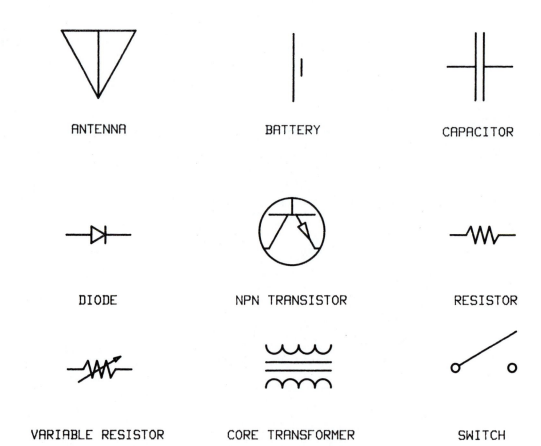

ANTENNA   BATTERY   CAPACITOR

DIODE   NPN TRANSISTOR   RESISTOR

VARIABLE RESISTOR   CORE TRANSFORMER   SWITCH

FIGURE 7-44 Draw these schematic symbols, and store as symbols or menu items.

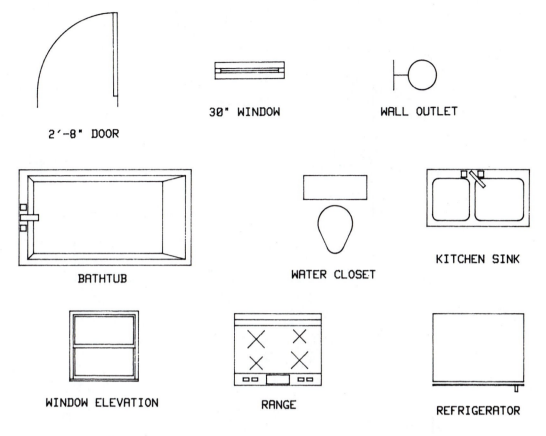

2′-8″ DOOR

30″ WINDOW

WALL OUTLET

BATHTUB

WATER CLOSET

KITCHEN SINK

WINDOW ELEVATION

RANGE

REFRIGERATOR

FIGURE 7-45 Draw these architectural symbols and store as symbols or menu items.

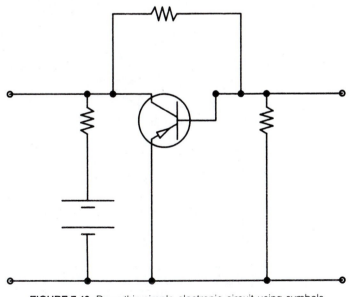

FIGURE 7-46 Draw this simple electronic circuit using symbols.

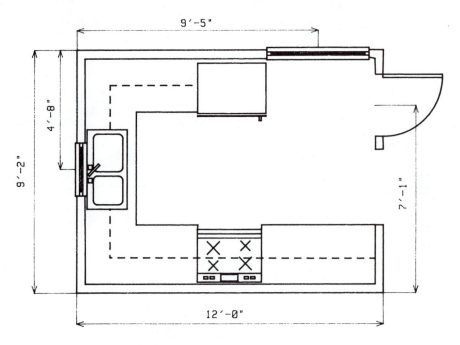

FIGURE 7-47 Draw this kitchen using symbols for the refrigerator, range, sink, and door.

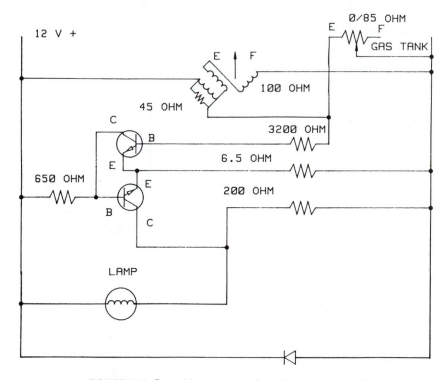

FIGURE 7-48 Draw this more complicated electronic circuit.

Basic Welding Symbols and Their Location Significance

Location Significance	Fillet	Plug or Slot	Spot or Projection	Seam	Back or Backing	Surfacing	Scarf for Brazed Joint	Flange
								Edge
Arrow Side					Groove weld symbol			
Other Side					Groove weld symbol	Not used		
Both Sides		Not used	Not used	Not used	Not used	Not used		Not used
No Arrow Side or Other Side Significance	Not used	Not used			Not used	Not used	Not used	Not used

Basic Welding Symbols and Their Location Significance

Flange	Groove							Location Significance
Corner	Square	V	Bevel	U	J	Flare-V	Flare-Bevel	
								Arrow Side
								Other Side
Not used								**Both Sides**
Not used		Not used	Not used	Not used	Not used	Not used	Not used	**No Arrow Side or Other Side Significance**

FIGURE 7-49 Draw each welding symbol shown and create a symbols library. *(Reproduced from AWS A.24-89 with permission of the American Welding Society.)*

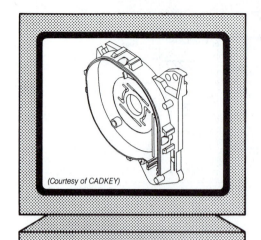

(Courtesy of CADKEY)

Chapter Eight

Basic Geometry— Drawing Circles, Lines, and Splines

The solutions to many design problems are best achieved through the use of basic geometric principles. Solutions are gained on the drafting board by using instruments and construction lines. CAD works in a similar way to solve problems. The major difference is that the solution to a problem is usually reached more quickly and involves less work because the computer does much of the construction and mathematics involved in solving the problem. For example, to draw a circle through three points with instruments involves a number of steps using construction lines, bisectors, and a high degree of accuracy. With CAD, the operator has only to identify the three points through which the circle is to be drawn, and the computer automatically calculates the tangencies and radius and draws the circle. This is only one example of how CAD can be used for basic geometric construction.

This chapter focuses on the geometric drawing exercises that can be performed by CAD. The major topics covered in this chapter include the drawing of circles, arcs, lines, construction lines, and splines.

PREREQUISITES

Before starting on this unit of instruction, you should be able to:

Draw

- lines and place points using CAD.

Use

- referenced and unreferenced points for drawing entities.
- existing points for drawing entities.

Place

- horizontal and vertical construction lines, use a grid, or input X-Y coordinates.

Make

- menu selections and draw simple orthographic views.

OBJECTIVES

After completing this chapter, you will be able to

- draw circles and arcs of specified radius and diameter.
- draw circles and arcs through two and three points, and tangent to entities.
- draw lines that are parallel, perpendicular, and tangent to other entities.
- draw construction lines that are parallel, perpendicular, and tangent to other drawing entities.
- draw fillets and chamfers.
- draw irregular curves using spline points.

MENU ITEMS IN THIS CHAPTER

1. **PARAL**—command used to draw lines parallel to existing lines.
2. **TYPED INPUT**—command used to tell the computer that you would like to use the keyboard to input values.
3. **PERP**—command used to draw lines perpendicular to existing lines.
4. **ANGLE**—command that will draw a line at an angle measured from an existing line and starting from a specified point.
5. **TANG**—command used to draw a line tangent to existing circles or arcs or to draw circles and arcs tangent to existing entities.
6. **SERIES**—subcommand used with PARAL to copy or parallel lines from a series of connected lines or arcs.
7. **MULT**—subcommand menu option used with PARAL to draw many lines parallel to an existing line of specified separation.
8. **CIRCLE**—command used to draw circles.
9. **ARC**—command used to draw arcs.
10. **C&R**—used with CIRCLE or ARC command to draw circles or arcs by locating the center point and keying-in or locating the radius.
11. **C&D**—used to locate circles and arcs by locating the center and locating or inputting the diameter.
12. **3 PTS**—used to locate circles or arcs by digitizing or inputting three points on the circumference.
13. **2 PTS**—menu item used to draw circles or arcs by digitizing or inputting two points on the circumference and inputting the radius.
14. **FILLET**—command used to draw a fillet or round on an existing corner and automatically clip the lines that extend beyond the arc.
15. **CHAMF**—command that will draw a chamfer on an existing corner and automatically clip the lines.
16. **SPLINE**—command used to draw irregular curves by digitizing points lying on the curve.

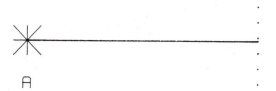

FIGURE 8-1 Drawing horizontal lines.

DRAWING LINES

When a drawing requires many lines, it is possible to simplify and speed-up the drawing process by using basic geometry drawing commands. For example, if you had to draw an object with many parallel lines it would be possible to use the PARAL drawing command to assist in drawing these lines. This section of the chapter shows how to use some of these commands for drawing lines. Many of these same commands are used to place construction lines on those systems that use construction lines as a means of creating a drawing.

Drawing Horizontal Lines

Sometimes, it is more convenient to place a line by identifying it as a horizontal or vertical line instead of placing it using coordinates, grids or construction lines. The following steps describe how to place horizontal lines on a drawing. Refer to Figure 8-1.

1. Select menu item LINE.
2. Select menu item HORZ.
3. Digitize one end point for the line (A).
4. Digitize the second end point of the line (B). A horizontal line will be drawn of a length equal to the horizontal distance between the two digitized points. Notice that the second point does not have to be at the location of the end point for the line. The perpendicular distance from point B determines the length.

Drawing Vertical Lines

Vertical lines can also be drawn without the use of grids, coordinates or construction lines by using the menu function VERT. This command works in a manner similar to HORZ lines. Refer to Figure 8-2 when studying the following steps for drawing vertical lines.

1. Select menu item LINE.
2. Select menu item VERT.

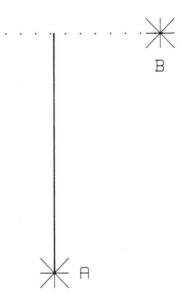

FIGURE 8-2 Drawing vertical lines.

3. Digitize the first end point of the line (A).
4. Digitize the second end point of the line (B). Notice that the perpendicular distance from the second digitized point determines the length of the vertical line.

Drawing Parallel Lines

Lines can be drawn parallel to a given line very easily with CAD. This is done by digitizing one end point of the parallel line, followed by digitizing the given line to which it is to be parallel. The approximate location of the end point of the parallel line is then digitized, causing a line to be drawn parallel to the given line starting from the first digitized point. TYPED INPUT can also be used to give an accurate position of the starting point of the line.

An example of drawing parallel lines is shown in Figure 8-3. Line A-B is the existing line.

1. Select menu item LINE.
2. Select menu item PARAL.
3. Digitize line A-B.
4. Digitize point C, which is the starting point of the parallel line.
5. Digitize point D. A line will be drawn parallel to and of the same length as line A-B. Notice that digitized point D does not have to be located at the actual end point of line C. Digitized point D only indicates the direction from which line C is drawn.

TYPED INPUT can also be used to draw parallel lines. This is accomplished by keying-in the starting point of the line relative to the origin,

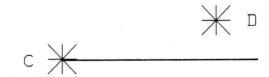

FIGURE 8-3 Drawing parallel lines.

and keying-in the length of the line. Figure 8-4 shows how this can be accomplished using CAD.

1. Select menu item LINE.
2. Select menu item PARAL.
3. Digitize line A-B.
4. Select menu item TYPED INPUT.
5. Key-in X4,YØ and RETURN.
6. Key-in L4 and RETURN. Line C-D will be drawn which is 4″ long and parallel to line A-B.

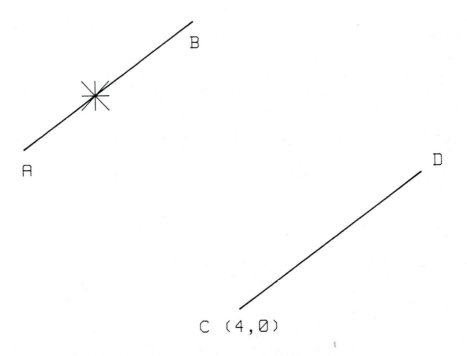

FIGURE 8-4 Drawing parallel lines using X-Y coordinates.

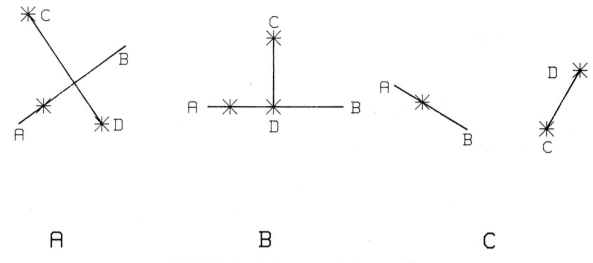

FIGURE 8-5 Drawing lines perpendicular to given lines.

Drawing Lines Perpendicular to Given Lines

Lines can also be drawn perpendicular to given lines using CAD. The procedures are very similar to those used in drawing parallel lines. Menu function PERP is used to activate this function. The end point of the perpendicular line is digitized. Then the given line is digitized, followed by the digitizing of the other end of the perpendicular line. Figure 8-5A shows the following steps.

1. Select menu item LINE.
2. Select menu item PERP.
3. Digitize point C, which is the start of the perpendicular line.
4. Digitize line A-B, which is the existing line that the new line is to be drawn perpendicular to.
5. Digitize point D, which is the end point of perpendicular line C-D.

The perpendicular line can be drawn anywhere on the screen in relation to the given line. Figures 8-5B and 8-5C show examples of this. A perpendicular line can also be drawn by using TYPED INPUT.

Drawing a Line at a Specified Angle from a Given Line

Lines can be drawn at a specified angle from a given line that starts from a specified point. This is accomplished by using menu item ANGLE. The angle is keyed-in, followed by the digitizing of the starting point of the line. The existing line is digitized, followed by locating the rough ending point of the new line. Figure 8-6 shows how to draw a line at an angle to an existing line, using the following steps.

1. Select menu item LINE.

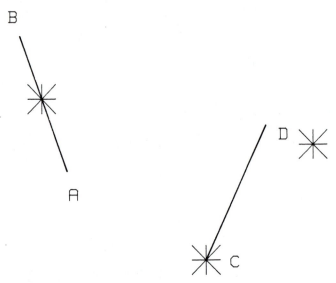

FIGURE 8-6 Drawing lines at a specified angle from an existing line.

2. Select menu item ANGLE.
3. A prompt will appear asking the operator to input the angle of the line to be drawn. Key-in 45 and RETURN.
4. Digitize the starting point of the new line (C).
5. Digitize the existing line.
6. Digitize the rough ending point of the new line (D). Line C-D is drawn at an angle of 45 degrees to line A-B.

Drawing Lines Tangent between Two Arcs or Circles

Lines can be drawn between two existing arcs or circles very easily with CAD. Menu function TANG is used to draw tangent lines. This is done by digitizing the two entities. The computer automatically draws a straight line that is tangent to the two entities. Figure 8-7 shows how a tangent line

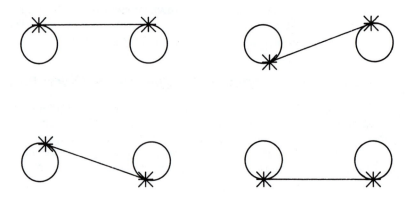

FIGURE 8-7 Drawing lines tangent to arcs or circles.

can be drawn between two full circles, depending on where the circles are digitized, using the following steps.

1. Select menu item LINE.
2. Select menu item TANG.
3. Digitize the first circle.
4. Digitize the second circle. A line will be drawn tangent to the two circles closest to the digitized points on the circle.

Drawing a Series of Parallel Lines

Some CAD systems have the capability of drawing multiple parallel lines. SERIES is the menu function used in this example. This function allows the user to copy or draw parallel lines from a series of connected lines or arcs. This can be a very effective function and one that can lead to some highly artistic results once the operator learns how to use it. Refer to Figure 8-8 for the following steps.

1. Select menu item LINE.
2. Select menu item PARAL.
3. Select menu item SERIES.
4. Digitize one end of the connected entities (1).
5. Digitize the other end of the entities (2).
6. The user can either digitize the new location for the parallel lines or choose TYPED INPUT and key-in the distance from the existing entities. A whole series of parallel entities can be created by digitizing more points or by keying-in multiple distances by separating with commas. In this example, multiple Vs are created by using TYPED INPUT and keying-in .25, .5, .75, 1, and so forth.

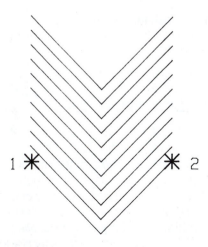

FIGURE 8-8 Drawing a series of parallel lines.

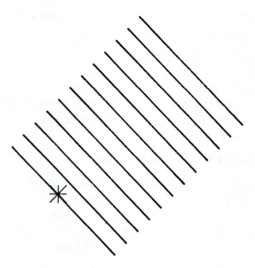

FIGURE 8-9 Drawing multiple parallel lines.

Drawing Multiple Parallel Lines

Another very useful and related function to the series parallel is the MULT function. This function allows the user to draw many lines parallel to a given line by keying-in the separation and the number of lines to be drawn. **Refer to Figure 8-9 to draw multiple parallel lines, using the following steps.**

1. Select menu item LINE.
2. Select menu item PARAL.
3. Select menu item MULT.
4. Digitize the line.
5. A prompt will ask for the separation between parallel lines. Key-in .25 and RETURN.
6. Another prompt will ask for the number of lines. Key-in the number 1Ø and RETURN. Ten lines of the same length will be drawn parallel to the given line, spaced .25″ apart.

DRAWING CIRCLES WITH CAD

Up to this point, lines are the only entity that has been explained. Obviously, most drawings require circles to be drawn. Using CAD, circles can be located in a variety of ways, adding to the flexibility of systems. Many methods can be used to place circles on drawings with CAD, because of the many different situations that an operator may encounter when drawing. The steps necessary to draw circles are explained in this section, thus allowing the user to create more complex drawings. Menu function CIRCLE is used to draw circles. Submenu commands are also added to allow the user a variety of options in drawing circles.

Drawing Circles of a Specified Radius and Known Center

One method of drawing circles is to use a known radius and the known location for the center point for the circle. The operator will use a menu item called C&R (Center & Radius) for this operation. The user then locates the center of the arc by digitizing or using TYPED INPUT. The radius is either input through the keyboard or digitized. The circle is then drawn about the center to the specified radius. Refer to Figure 8-10A for drawing circles using the C&R command.

1. Select menu item CIRCLE.
2. Select menu item C&R.
3. Digitize the center point (A).
4. Digitize a point on the circumference of the circle (B). The circle is drawn using A as the center and A-B as the radius.

Refer to Figure 8-10B and use the following steps with C&R for digitizing the center and inputting the radius.

1. Select menu item CIRCLE.
2. Select menu item C&R.
3. Digitize the center point.

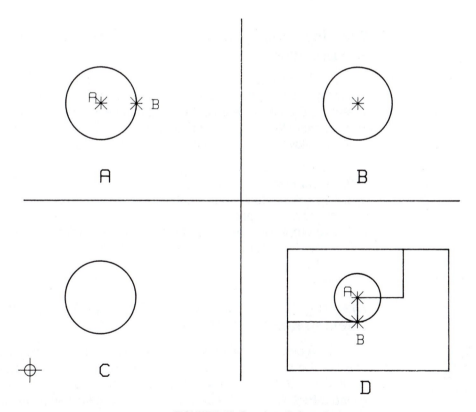

FIGURE 8-10 Drawing circles.

4. Select menu item TYPED INPUT.
5. Key-in R.75 and RETURN to draw a circle having a .75″ radius.

Refer to Figure 8-10C and use the following steps with C&R and TYPED INPUT to key-in the positions of the center and the radius.

1. Select menu item CIRCLE.
2. Select menu item C&R.
3. Select menu item TYPED INPUT.
4. Key-in the position of the center point from the origin, such as X2,Y2 and RETURN.
5. Key-in a point on the circumference, such as X1,Y2, and RETURN to draw a 1″ circle.

When using digitized points to draw circles, the points can be existing points such as drawing entities, grid points or construction lines. The following steps and Figure 8-10D show how existing points are used to draw a circle.

1. Select menu item CIRCLE.
2. Select menu item C&R.
3. Digitize existing point A for the center of the circle.
4. Digitize existing point B for a point on the circumference of the circle.

Drawing Circles by Locating the Center and Inputting the Diameter

A circle can also be drawn by locating the center of the circle and inputting the diameter of the circle through a menu function called C&D (Center & Diameter). The center point can be located by digitizing or inputting the exact location through coordinates. Refer again to Figure 8-10B and use the following steps to learn how to use the C&D function.

1. Select menu item CIRCLE.
2. Select menu item C&D.
3. Digitize the center point of the circle or use TYPED INPUT to key-in the coordinate location from the origin.
4. A prompt will appear asking to input the desired diameter. The user keys-in the diameter followed by RETURN to draw the circle as shown in the figure.

Drawing Circles by Locating Three Points on the Circumference

A circle can be drawn by locating three points along the circumference of the circle. These points can be digitized from existing points or located by using TYPED INPUT and coordinates. This menu function is called 3 PTS. Refer to Figure 8-11 and the following steps to use 3 PTS for drawing a circle.

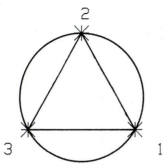

FIGURE 8-11 Drawing a circle through three points.

1. Select menu item CIRCLE.
2. Select menu item 3 PTS.
3. Digitize in order the three points to be located on the circumference of the circle. In this example, the three points of the triangle are digitized. TYPED INPUT could also be used by keying-in the coordinate values for each point.

Drawing Circles by Locating Two Points on the Circumference

Circles can be drawn by digitizing or keying-in the location of the two points that are to lie on the circumference and inputting the radius. This menu function is called 2 PTS. Refer to Figure 8-12 and the following steps to use 2 PTS.

1. Select menu item CIRCLE.
2. Select menu item 2 PTS.
3. A prompt will appear asking to input the desired radius. The operator keys-in the radius followed by RETURN.
4. Digitize point A and point B to draw the circle between the two existing end points of the lines.

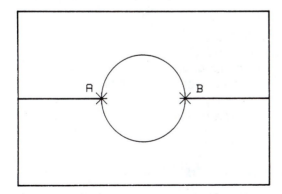

FIGURE 8-12 Drawing a circle through two points.

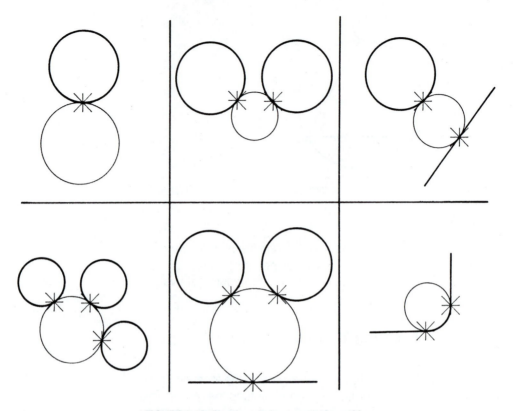

FIGURE 8-13 Drawing circles tangent to entities.

Drawing Circles Using Tangencies and a Specified Radius

Circles can be located tangent to existing entities by digitizing the entities and keying-in the radius. The menu function used for drawing circles in this manner is TANG. Figure 8-13 shows some of the different situations that can be used with this function. The following steps show how this function is used to draw circles.

1. Select menu item CIRCLE.
2. Select menu item TANG.
3. A prompt will appear asking to input the desired radius. The operator keys-in the value followed by RETURN.
4. The operator then digitizes the entity or entities to which the circle is to be tangent followed by RETURN and the computer draws the circle.

DRAWING ARCS USING CAD

The drawing of arcs with CAD has many similarities to the drawing of circles. As with circles, there are many different ways to produce arcs.

This section explains how arcs are drawn. Menu function ARC is used to draw arcs with submenu commands to allow the user a variety of options.

Drawing Arcs of a Specified Radius and Known Center

Arcs can be drawn by inputting the radius, identifying the center point, and locating the end points of the arc. These points can be digitized using existing points or located with TYPED INPUT. Menu item C&R (Center & Radius) is used to draw arcs in this manner. Refer to Figure 8-14A and the following steps to draw arcs using menu item C&R.

1. Select menu item ARC.
2. Select menu item C&R.
3. A prompt appears asking for the radius of the arc. The operator keys-in the radius followed by RETURN.
4. Digitize the location of the center point (A).
5. Digitize the start of the arc (B).
6. Digitize the end of the arc (C) to produce arc B-C.

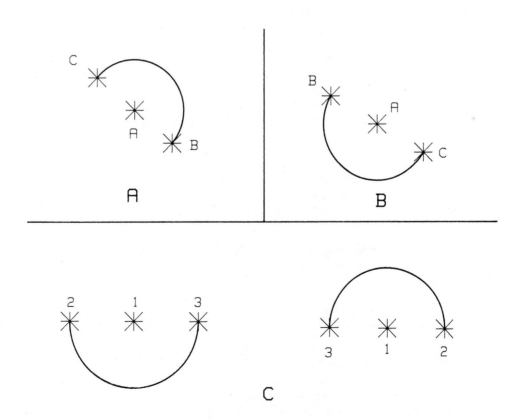

FIGURE 8-14 Drawing arcs.

Drawing Arcs of a Specified Diameter and Known Center

Arcs can also be drawn by keying-in the diameter, locating the center point, and locating the end points of the arc. This method uses menu function C&D (Center & Diameter). Refer to Figure 8-14B to draw an arc using C&D.

1. Select menu item ARC.
2. Select menu item C&D.
3. A prompt will appear asking the operator to input the diameter of the arc. Key-in the desired diameter followed by RETURN.
4. Digitize the center point of the arc (A).
5. Digitize the start of the arc (B).
6. Digitize the end of the arc (C) to draw arc B-C.

NOTE: The direction in which the arc is drawn is determined by the specific type of CAD system being used. Some CAD systems will draw only in one direction, such as clockwise or counterclockwise. Other systems allow the operator to choose the direction of drawing through a menu function. Some systems will draw in the direction in which the points were digitized. An example of this is shown in Figure 8-14C. This is also an example of a system that draws circles and arcs in a counter-clockwise direction only.

Drawing Arcs by Digitizing Three Points

Arcs can be drawn by digitizing three points located on the circumference of the arc. These points can be digitized by using existing points or by using TYPED INPUT. The menu function used for this operation is 3 PTS. Refer to Figure 8-15A while using the following steps.

1. Select menu item ARC.
2. Select menu item 3 PTS.
3. Digitize the start of the arc (A).
4. Digitize a point on the circumference of the arc (B).
5. Digitize the end point of the arc (C) to draw arc A-B-C.

Drawing Arcs by Inputting the Radius and Digitizing the End Points

Arcs can be drawn by inputting the radius of the arc and digitizing the start and end of the arc. Menu function 2 PTS is used for this drawing exercise. Refer to Figure 8-15B and the following steps.

1. Select menu item ARC.
2. Select menu item 2 PTS.
3. A prompt appears asking for the radius of the arc. The operator keys-in the radius followed by RETURN.
4. Digitize the start of the arc (A).
5. Digitize the end of the arc (B) to draw arc A-B.

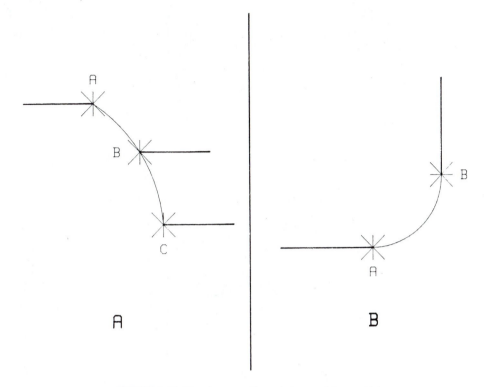

FIGURE 8-15 Drawing arcs through two and three points.

Drawing Arcs Using a Tangent Entity and a Specified Radius

Arcs can also be drawn tangent to an entity or entities by digitizing the entities and inputting the desired radius. Menu function TANG is used for this arc drawing subroutine. Figure 8-16 shows some of the different ways that this command can be used to draw arcs. The following steps show how this command might be used to draw an arc.

1. Select menu item ARC.
2. Select menu item TANG.
3. A prompt will appear asking the operator to input the desired radius. Key-in the radius followed by RETURN.
4. Digitize the start of the arc and the points of tangency.
5. Digitize the end of the arc. An arc of the specified radius will be drawn between and tangent to the digitized points.

AUTOMATIC FILLETING WITH CAD

It is often necessary to round square corners on drawings. These arcs are called rounds and fillets. Drawing fillets between two existing lines with CAD can be done easily by using the ARC command. However, once

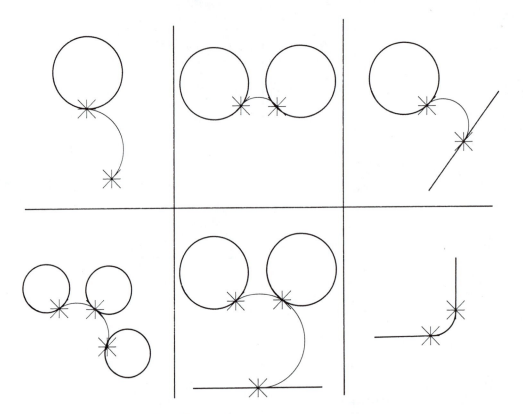

FIGURE 8-16 Drawing arcs tangent to entities.

the arc is drawn, part of the existing lines must be erased to leave only the fillet or round as shown in Figure 8-17A. Using a function called CLIP, as explained in Chapter Six, a user can delete the portion of the lines not needed. However, some systems have a much faster method for drawing fillets and rounds.

Many CAD systems have an automatic filleting routine that will place an arc at a corner and automatically clip the lines. Having this function speeds up drawing arcs at corners by eliminating the need to use the CLIP command. The menu command FILLET is used to describe this function. To use this command, some systems require the operator to input the desired radius of the fillet and locate the corner by digitizing the lines. Other systems have a default value such as ⅛″ that can be changed to your specifications. The example used here has the computer automatically prompt the operator to input the desired radius. Figure 8-17B and the following steps show how the FILLET command is used to draw fillets or rounds at existing corners.

1. Select menu item FILLET.
2. A prompt will appear asking the operator to input the radius. Key-in the radius followed by RETURN.
3. Digitize the two lines that make up the corner. The arc will be drawn and the lines automatically clipped as shown in the figure.

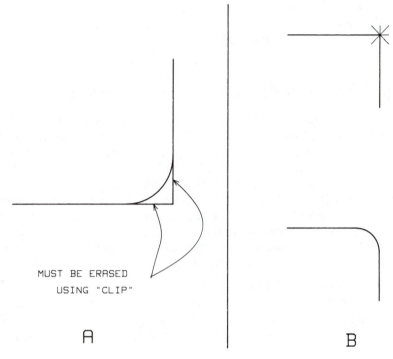

MUST BE ERASED
USING "CLIP"

A   B

FIGURE 8-17 Drawing fillets.

AUTOMATIC CHAMFERING WITH CAD

Another time-saving command that is available on some CAD systems is a function used to chamfer corners. This function automatically places a chamfer on a corner and clips the lines. The type of CAD system determines how the chamfer is placed. Either a prompt appears asking for the angle and size, or a default value may be used. The example used here has a default value of ¼″ and 45 degrees. The menu function used is CHAMF. Figure 8-18 and the following steps explain how this function is used.

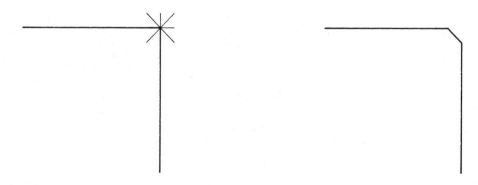

FIGURE 8-18 Drawing chamfers.

1. Select menu item CHAMF.
2. Digitize the corner. The corner will be clipped automatically and a ¼″ x 45 degree chamfer will be drawn.

DRAWING IRREGULAR CURVES USING CAD

Sometimes, it is necessary to draw curves that are not arcs or circles. These curves are used in drawing cams, sheet metal layouts, charts, and so forth. An irregular or French curve or a spline are the instruments used to draw such curved surfaces in traditional drafting. CAD systems have a function commonly called SPLINE or BEZIER that can be used for this type of drawing. With this function the operator identifies the points through which the irregular curve is to be drawn, and the system automatically calculates the curve and draws the smooth, irregular shape through the digitized points. An example of how this function is used is shown in Figure 8-19, as explained in the following steps.

1. Select menu item SPLINE.
2. Digitize the intersecting points for the cam, moving in a clockwise or counterclockwise direction. When the last point is digitized, press RETURN. A smooth curve will be drawn through each digitized point.

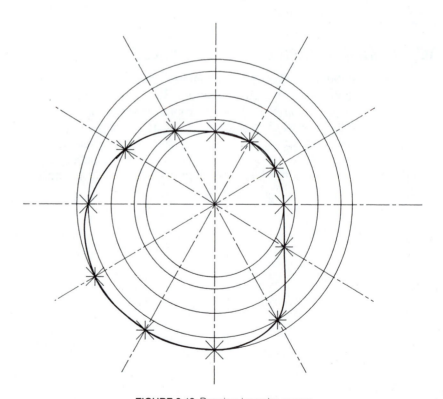

FIGURE 8-19 Drawing irregular curves.

SUMMARY

This chapter shows how to draw circles and arcs in a variety of ways. Some basic geometry practices are reviewed, leading the user to discover how fast and simple geometric construction can be with CAD as compared to hand drawing. The information in this chapter provides the operator with the knowledge necessary to complete more complex drawings with CAD. The following chapters build upon this knowledge, add more functions, and show the operator faster and more efficient methods of drawing.

Chapter Eight REVIEW

1. List the commands used on your CAD system to draw parallel and perpendicular lines.
2. List the commands used on your CAD system to draw horizontal and vertical lines.
3. List the different commands used on your CAD system to draw circles and arcs.
4. Determine if your CAD system has a fillet and chamfer function and list the steps to place them on a drawing.
5. List the steps used to draw irregular curves with your CAD system.

Chapter Eight DRAWING EXERCISES

1. Draw a series of circles and arcs using the different circle and arc commands on your system.
2. Using the circles and arcs drawn in question 1, draw tangent lines between some of the arcs or circles.
3. Draw a series of lines using Figure 8-20 as an example. Now repeat that series of lines 1'' from and parallel to the existing lines by using a function such as SERIES, as described in this chapter.

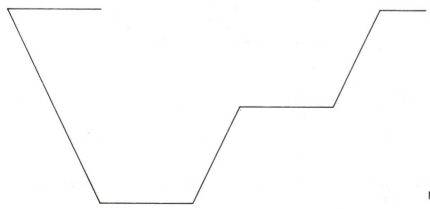

FIGURE 8-20 Drawing a series of lines.

4. Draw five 2″ horizontal lines spaced ½″ apart by using a command such as MULT, as described in this chapter.

5. Draw an equilateral triangle having 3″ sides, then circumscribe a circle around the triangle by using a command such as 3 PTS, as described in this chapter.

6. Draw a 2″ x 4″ rectangle, then use the FILLET command to draw ⅛″ rounds on each corner.

7. Draw a 2″ x 4″ rectangle, then use the CHAMF command to draw ¼″ chamfers on each corner.

8. From the exercises that follow, draw those selected by your instructor.

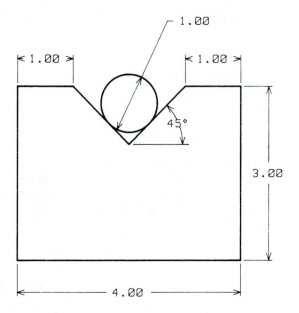

V-BLOCK

FIGURE 8-21 Draw the V-Block and add a 1″ diameter circle tangent to the V-groove. Do not dimension.

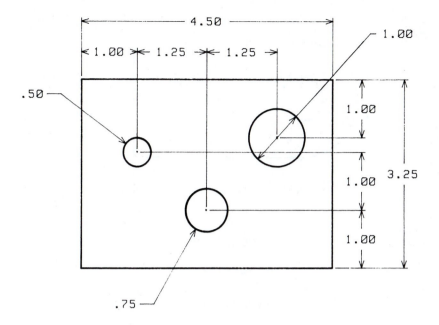

FACE PLATE

FIGURE 8-22 Draw the Face Plate. Do not dimension.

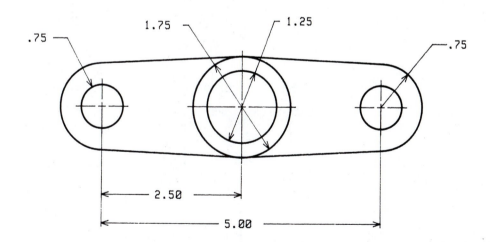

SPACER

FIGURE 8-23 Draw the Spacer. Do not dimension.

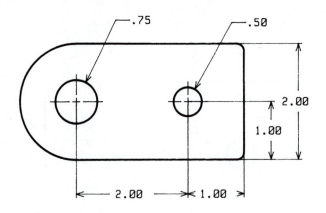

ALL ROUNDS .25"

END BRACKET

FIGURE 8-24 Draw the End Bracket. Do not dimension.

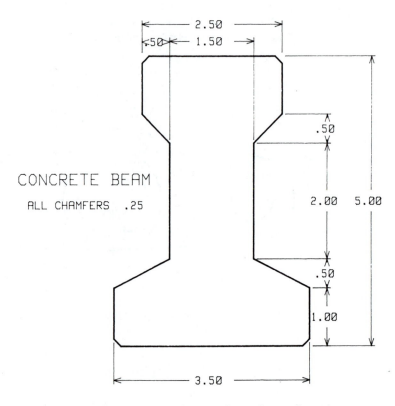

CONCRETE BEAM

ALL CHAMFERS .25

FIGURE 8-25 Draw the Concrete Beam. Do not dimension.

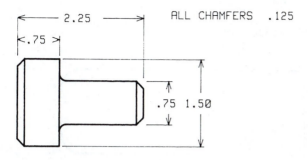

ALL FILLETS   .125

ALL CHAMFERS   .125

LINK PIN

FIGURE 8-26 Draw the Link Pin. Do not dimension.

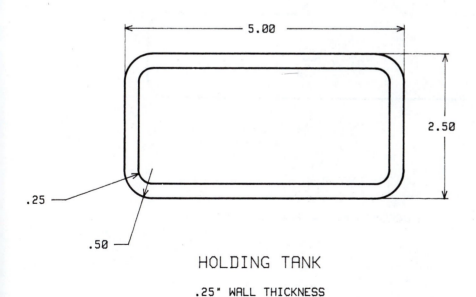

HOLDING TANK

.25" WALL THICKNESS

FIGURE 8-27 Draw the Holding Tank. Do not dimension.

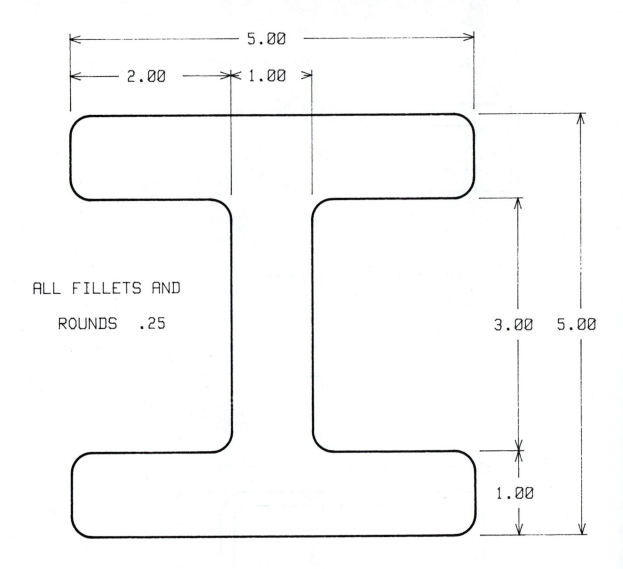

5.00

2.00

1.00

ALL FILLETS AND

ROUNDS .25

3.00   5.00

1.00

I BEAM

FIGURE 8-28 Draw the I Beam. Do not dimension.

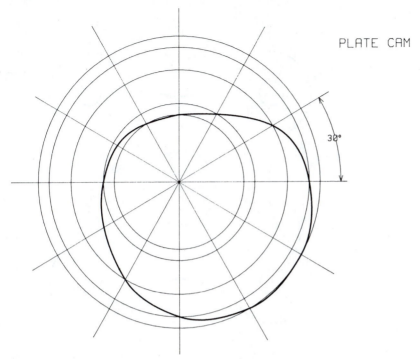

PLATE CAM

30°

FIGURE 8-29 Draw the Plate Cam. Do not dimension. Circle diameters are: 3″, 3.5″, 5″, 6″, and 6.5″.

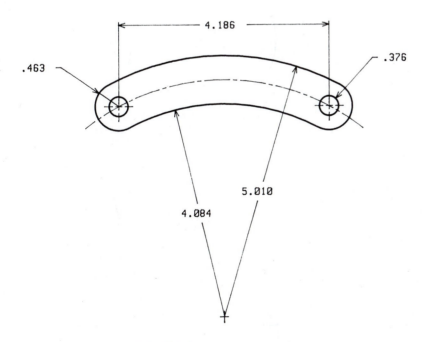

4.186

.463

.376

5.010

4.084

FIGURE 8-30 Draw the latch plate.

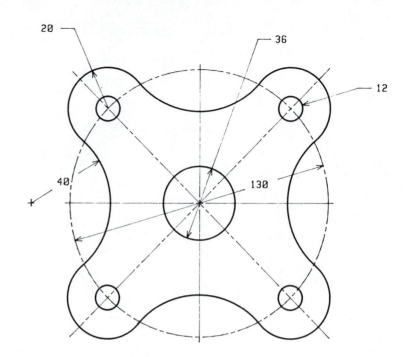

FIGURE 8-31 Draw the multispacer (metric).

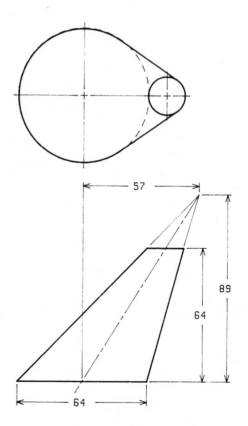

FIGURE 8-32 Develop the truncated cone (metric).

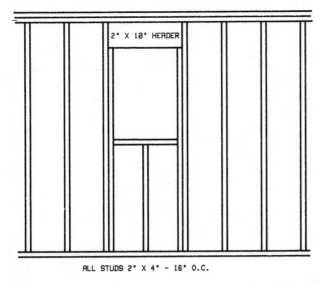

ALL STUDS 2" X 4" - 16" O.C.

FIGURE 8-33 Draw the wall detail, scaling to fit an 8-½" x 11" or 11" x 17" sheet.

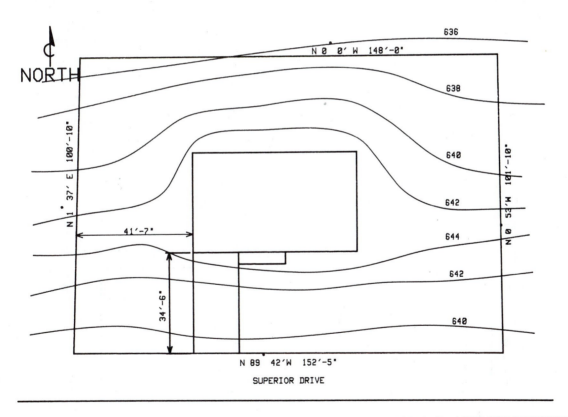

LOT NO. 7, WHITE PINE SUBDIVISION
FAITHORN COUNTY, MICHIGAN

FIGURE 8-34 Draw the plot plan, estimating unknown dimensions. Scale to fit an 8-½" x 11" or 11" x 17" sheet.

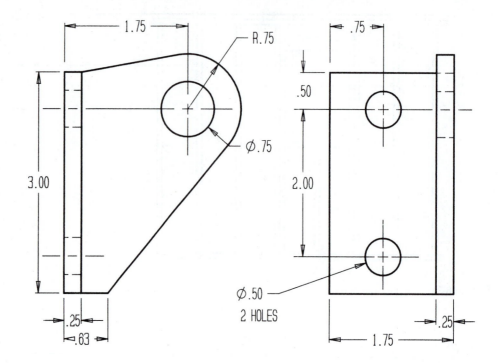

FIGURE 8-35 Pivot Bracket.

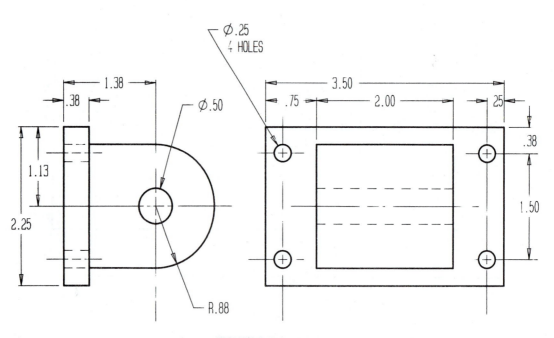

FIGURE 8-36 Tie Brace.

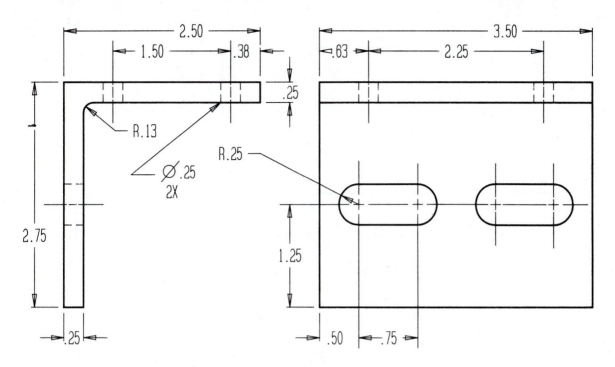

FIGURE 8-37 Contact.

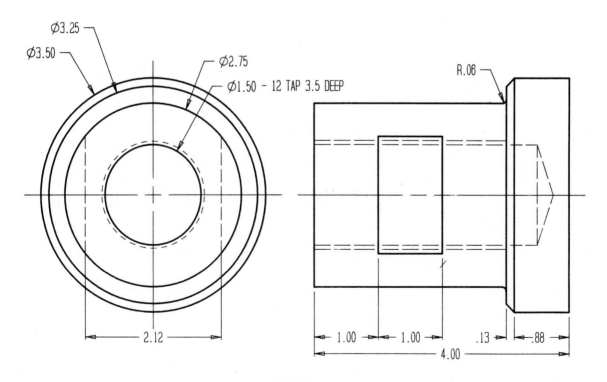

FIGURE 8-38 Brass Plug.

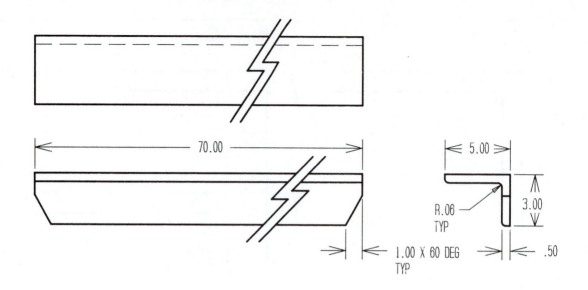

FIGURE 8-39 Tie Brace.

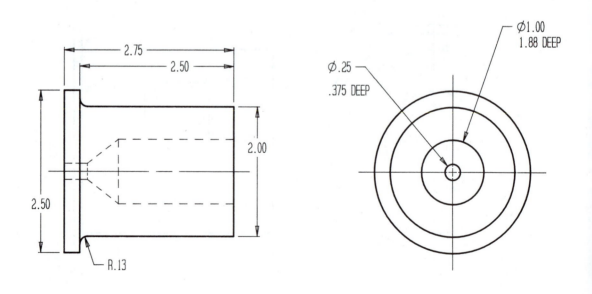

FIGURE 8-40 Cylinder.

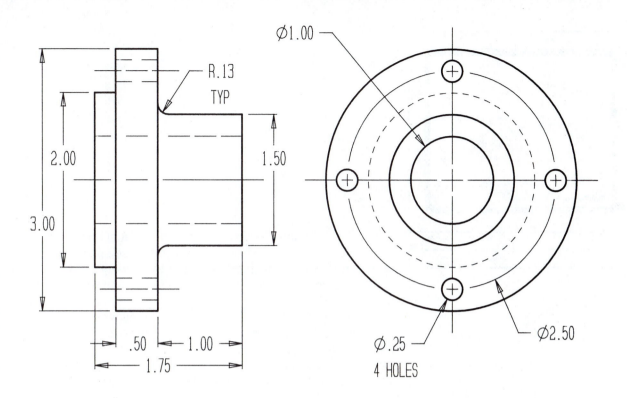

FIGURE 8-41 Shaft Support.

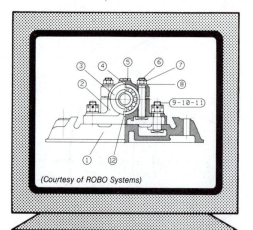

(Courtesy of ROBO Systems)

Chapter Nine

Sectional Views, Labels, Dimensions, and Special Functions

This chapter moves away from the actual drawing of an object with lines, arcs and circles. To be discussed now are some of the special features that are added to views of an object to clarify features, place labels, and add dimensioning. One method used to clarify drawings is through the use of cutaways or sectional views. In conventional drafting practices, an object is sectioned to reveal interior features by adding section lines to represent surfaces cut by the cutting plane and eliminating hidden lines. These lines must be drawn one at a time by hand using traditional tools. With CAD, the area is defined, the crosshatching pattern is specified or chosen from a menu of different patterns, and the area is automatically filled with the chosen pattern.

Annotation is added to a drawing in the form of labels and dimensions. Lettering is one of the most time-consuming and tedious tasks that a drafter can encounter on a drawing. CAD alleviates this tedium because placing labels is simply a matter of typing-in the characters from the keyboard or menu. Labels are placed by digitizing the position and keying-in the label. Labels come in many different styles, sizes, and slants, all of which are controlled by the CAD operator.

Dimensioning is another type of annotation that is a very tedious job for the drafter. CAD makes dimensioning very simple and fast. For linear dimensions, the two points to be dimensioned are digitized. The position of the dimension figure is then digitized. The computer automatically calculates the dimension, and then places dimension lines, extension lines, arrowheads, and the dimension figure on the drawing.

Other functions covered in this chapter include the placing of tolerance figures, and geometric tolerances and symbols. Two other areas related to dimensioning reviewed here are calculating an area and measuring distances and angles. At the conclusion of this chapter, you should be able to create a complete detail drawing of a part. You should also become more proficient at doing such simple tasks as drawing lines. You may also become more confused as additional commands are added. This will be overcome as more time is spent on the CAD system. Thinking over the steps you need to follow before starting on a drawing will help to prevent frustration and time lost in drawing an object.

PREREQUISITES

Before starting on this unit of instruction, you should be able to:

Draw

- simple orthographic views of an object.
- circles and arcs.
- entities using basic geometric principles.

Demonstrate

- a knowledge of sectional views.
- a knowledge of accepted dimensioning techniques.
- a knowledge of placing tolerances on drawings.
- a knowledge of geometric tolerancing symbology.

OBJECTIVES

After completing this chapter, you will be able to

- define an area and place section lines to your specifications.
- place labels to your specifications.
- dimension an object completely.
- measure entities on a drawing.
- calculate a defined area on a drawing.
- complete a detailed drawing of a simple part.
- place tolerances using CAD.
- place geometric tolerances and symbols using CAD.

MENU ITEMS IN THIS CHAPTER

1. **HATCH**—command used to place section lines in a specified area.
2. **DEFINE**—submenu item used with HATCH to allow the user the option of defining the angle, spacing, and line type to be used on a pattern fill.
3. **TEXT**—command used to place labels on the drawing.
4. **HGT**—submenu command used with TEXT to change the text height.
5. **SLANT**—submenu item used with TEXT to change the angle of the characters.
6. **ANGLE**—submenu item used with TEXT to change the angle at which the label will be drawn.
7. **LINE WEIGHT**—command used to change the line weight used by the system.

8. **JUSTIFY**—submenu item used with TEXT to change the location point when digitizing the position of the label.
9. **MULT**—submenu command used with TEXT to place more than one line of text by automatically spacing each line equally.
10. **DIMEN**—command used to place linear dimensions on a drawing.
11. **DIA**—command used to place dimensions on circles.
12. **RADIAL**—command used to place dimensions on arcs.
13. **ANGULAR**—command used to place dimensions on angles.
14. **UNITS**—command used to change the units used when dimensioning, such as millimeters or inches.
15. **LEADERS**—command used to place leader lines and arrowheads used as pointers.
16. **DIM HGT**—command used to change the height of the text when the dimension text is placed using the dimensioning commands.
17. **TOLER**—command used to specify a tolerance figure that is to be placed after the dimension figure when using the dimensioning commands.
18. **GEO TOL**—command used to place geometric tolerance symbols on a drawing.
19. **MEAS DIST**—command used to measure a linear dimension.
20. **MEAS ANGLE**—command used to measure the angle between two intersecting lines.
21. **CALC AREA**—command used to calculate the area within a defined boundary.
22. **PERIM**—command used to calculate the actual or projected lengths of a single entity or a group of entities.
23. **CENTROID**—command to determine the center of gravity of a closed figure.
24. **MOMENT**—command that calculates the moment of inertia.
25. **VERIFY**—command that determines various values of an entity such as level, color, and so forth.

DRAWING SECTIONAL VIEWS

Sectional views are used to reveal interior features not clearly seen using hidden lines. Many different types of sectional views are used, such as full sections, half sections, offset sections, and so forth. All sectional views require the drawing of section lines to represent surfaces cut by the cutting plane. These surfaces can be defined by identifying the entities making up the boundary of the surface. By identifying the boundary, the computer can automatically place section lines. The spacing, angle, line type, and material symbol can all be controlled by the operator.

The steps to follow in placing section lines depend on the boundary of the surface and the type of section symbol used. Two methods of defining the boundary are shown here: automatic and manual. Two methods of selecting the crosshatching symbol are used: a user-defined crosshatching pattern and a predefined crosshatching symbol. Various area boundaries are also used to demonstrate the different steps involved in placing crosshatching. Menu commands used for this function are usually "fill," "hatch," "xhatch" or just a pattern fill drawn on the menu. HATCH is the menu command used in this text.

CROSSHATCHING SYMBOLS

Most CAD systems have a variety of crosshatching symbols available to the user. This is done so that the user does not have to spend time defining the type of crosshatching necessary for a particular fill pattern. Figure 9-1 shows some common crosshatching patterns found in CAD. These patterns may be drawn on the menu or represented by a number. The user chooses the pattern desired by digitizing the pattern or the number representing the pattern on the menu, and keying-in the angle and spacing if different from the default values displayed in the prompt.

User-defined Crosshatching

Sometimes, the standard pattern symbols on the menu are not the type needed for a specific sectional drawing. When this occurs, the user has the option of defining the type of symbol needed. The user does this by inputting the line type, spacing, and angle. By so doing, the user has the ability to define the type of crosshatching needed for most pattern fills. Menu item DEFINE is used with HATCH to give the user the option of defining the

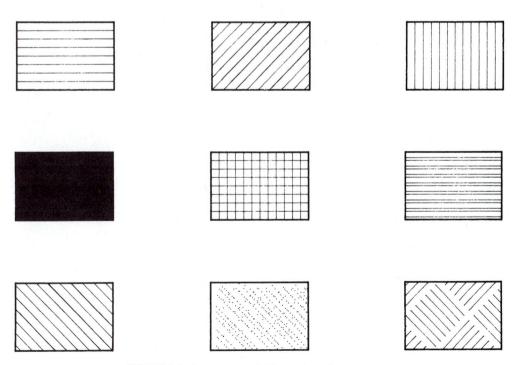

FIGURE 9-1 Some typical CAD crosshatching patterns.

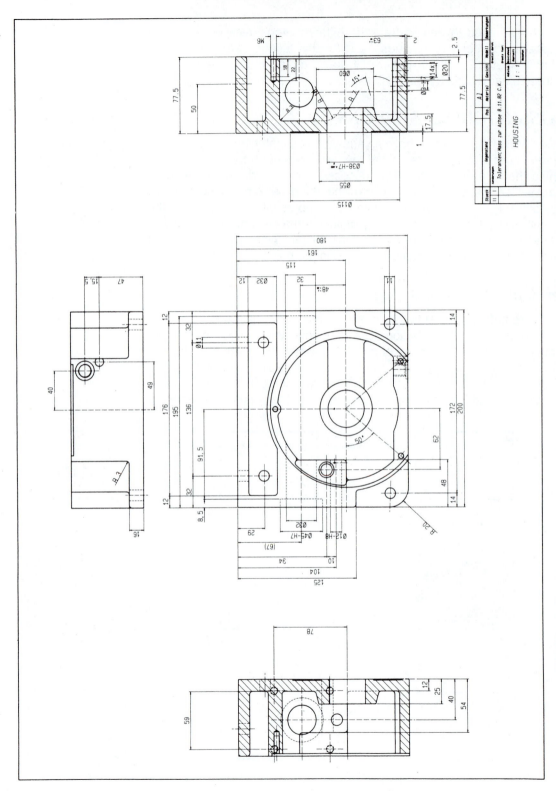

FIGURE 9-2 Fill patterns represent sectional views of a housing produced with CAD *(Courtesy of Hewlett-Packard)*

pattern fill. Figure 9-2 shows a drawing produced with CAD having fill patterns to represent sectional views.

Defining the Fill Area Automatically

Automatic boundary definition is common on many CAD systems. This is accomplished by digitizing one line on the boundary of the area. The computer will automatically define the boundaries around the surface to be sectioned. This is accomplished in a variety of ways, depending upon the particular CAD system being used. One method used is to have a separate menu item for automatic crosshatching. Another method used is to have the system automatically prompt the user for automatic or manual boundary definition after the "hatch" or "fill" command is selected. The latter method is used in this text. Refer to Figure 8-3 and the following steps to fill the object with crosshatching.

1. Select menu item HATCH.
2. The system prompts for "automatic" or "manual" area definition. For this example, select "automatic."
3. Select the pattern fill needed for the section by digitizing the fill pattern on the menu. Select the angle and spacing desired for the fill.
4. Digitize one of the lines of the object. The computer will automatically locate all the lines that make up the area. Usually a small "tick" (square, circle, slash) will be drawn on each entity or corner of the object as the computer searches the lines that make up the surface area. If the computer comes to a corner with more than two lines, it will stop and prompt the user to digitize the line to be included in the fill border. In this example, the computer defines the area without user input.
5. The computer then fills the area with the crosshatching pattern selected.

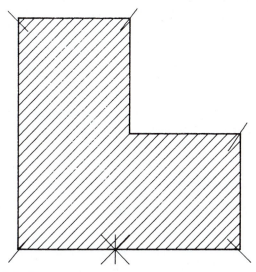

FIGURE 9-3 Filling a defined area with crosshatching.

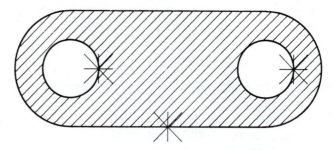

FIGURE 9-4 Unfilled details in a filled area.

Unfilled Details in a Filled Area

When filling an area, it is not uncommon to find a hole or some other feature that does not require section lines to be drawn because the surface has not been touched by the cutting plane. This area without section lines may lie within the area to be crosshatched, as shown in Figure 9-4, and it is sometimes referred to as a *negative area*. When this occurs, you must be able to communicate to the computer which areas are to be filled and which areas are not to be filled. This is done by automatically or manually defining the area to be filled, then digitizing the area that is to be left without crosshatching. Refer to Figure 9-4 and the following steps to fill an area without filling details in that area.

1. Select menu item HATCH.
2. The system prompts for automatic or manual area definition. Select "automatic."
3. Select the pattern fill desired from the menu. Select the angle and spacing desired for the fill.
4. Digitize a line or arc on the drawing.
5. Digitize each circle. This indicates to the computer that the areas within the circles are not to be filled. The computer will fill the area defined without filling the circles.

Defining the Fill Area Manually

There are times when the area to be filled would be difficult to do automatically because of the many lines, arcs and circles within the area. This is when it is more convenient to use the manual method of defining areas to be filled. When "manual" is selected, you must digitize each entity

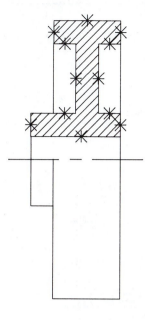

FIGURE 9-5 Defining fill areas manually.

that makes up the boundary. Refer to Figure 9-5 and the following steps for defining fill areas using the manual mode.

1. Select menu item HATCH.
2. The system prompts for "automatic" or "manual" area definition. Select "manual."
3. Select the pattern fill desired from the menu. Select the spacing and angle desired for the fill pattern.
4. Digitize each entity that borders the area to be filled. A small "tick" or slash will be drawn on the entity or at the corner of each entity as it is digitized. Complete the area definition by digitizing the first entity again and RETURN. The area defined will be filled with the pattern selected.

Filling an Area with User-defined Crosshatching

Because the operator has the option of selecting the line type, angle, and spacing of each fill pattern on the menu, most fill pattern requirements are met using menu fill patterns. Occasionally, there are times when you must produce your own fill patterns. Most CAD systems allow users to create their own pattern fills. The pattern fill is created by drawing the fill pattern and adding it to the menu or making it a symbol. When that particular pattern fill is needed, the user selects the user-defined fill from the menu or calls up from memory the pattern fill stored as a symbol and locates it in the defined area.

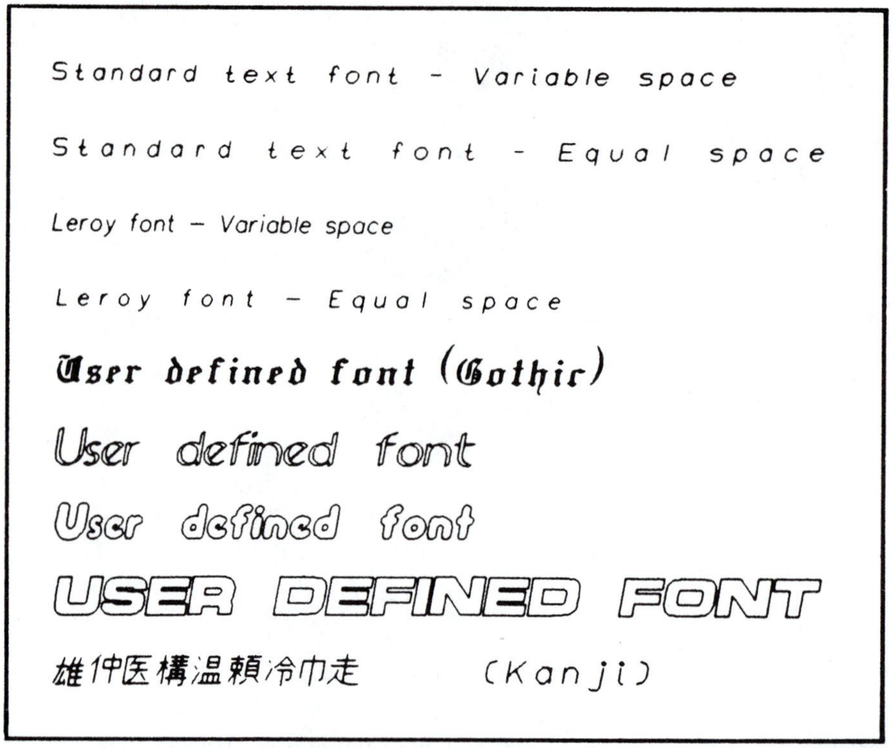

Standard text font - Variable space

Standard text font - Equal space

Leroy font - Variable space

Leroy font - Equal space

User defined font (Gothic)

User defined font

User defined font

USER DEFINED FONT

雄仲医構温頼冷巾走   (Kanji)

FIGURE 9-6 Typical text font available with CAD. *(Courtesy of Calma Company)*

PLACING LABELS ON THE DRAWING

Placing text using CAD is one of the major time-saving tasks performed by the computer. Compared to placing text by hand, CAD is many times more efficient and much neater. Most systems have more than one type of text font (style) from which to choose. Figure 9-6 shows the text fonts available to the user of one manufacturer's system. The type of text font used is selected at the same time the TEXT command is implemented by the user. Many variations for labels can be controlled by the user besides the lettering style. These include text height, character slant, angle of the text line, text justification when placing, thickness of text lines, and single or multiple lines.

Most systems have default values that are active when the TEXT command is chosen. Typical default values are .125″ for height, Ø degrees slant and angle, justified left, standard line thickness, standard lettering style, and a single line of text. These default values can be changed by selecting the menu item for each variable. In this text, menu items HGT, SLANT, ANGLE, LINE WEIGHT, and JUSTIFY are the commands used to change these default values. Figure 9-7 shows examples of variations in text used with labels.

The placing of text on the drawing is a very easy task. The user selects the menu item for text and the text font to be used. Any default

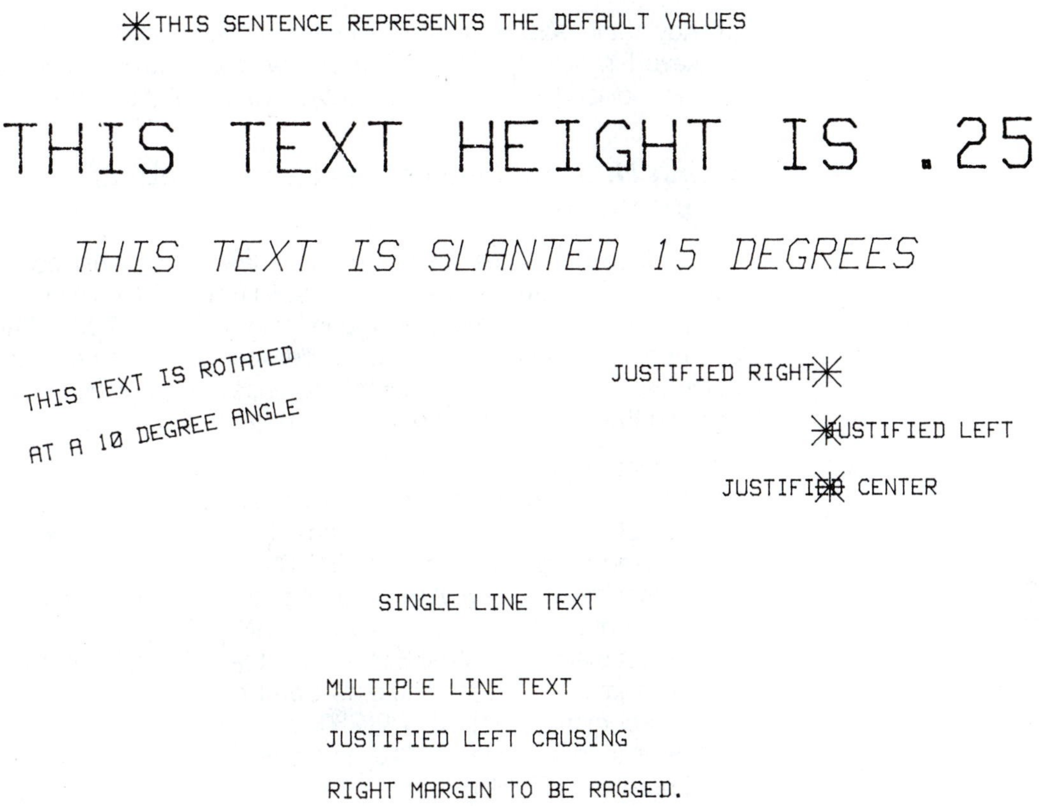

✳THIS SENTENCE REPRESENTS THE DEFAULT VALUES

THIS TEXT HEIGHT IS .25

THIS TEXT IS SLANTED 15 DEGREES

THIS TEXT IS ROTATED
AT A 10 DEGREE ANGLE

JUSTIFIED RIGHT✳

✳JUSTIFIED LEFT

JUSTIFI✳ CENTER

SINGLE LINE TEXT

MULTIPLE LINE TEXT

JUSTIFIED LEFT CAUSING

RIGHT MARGIN TO BE RAGGED.

FIGURE 9-7 Example lines of text showing variations in size and style.

values to be changed are done by picking the menu item and keying-in the new value. The starting position for the label is then digitized if the label is justified left. If the label is justified center the digitized point will cause the label to be centered about that point. If the label is justified right the digitized point will be the end of the text line. Examples of the different justification points are shown in Figure 9-7. The label is then keyed-in using the computer keyboard or the menu if letters are included, followed by RETURN. The label is then drawn onto the screen in the size and style selected by the user.

Placing a Single-line Label

This example shows how to place a single-line label using the default values. Placing single-line text using default values is easily done because the operator has only to type-in the label and digitize the location of the label on the screen. Refer to the top line of text in Figure 9-7 and the following steps for single-line labels.

1. Select menu item TEXT.
2. Digitize the position for the start of the label.

3. Key-in the label using the keyboard or menu. After the label has been keyed-in, select RETURN. The label will be drawn on screen starting at the digitized point using the default values of the system.

Placing Multiple Lines of Text and Changing Default Values

There are times when a multiple line of text must be used on a drawing for long notes or to explain surface finishes. When multiple-line text is chosen, each line of text is keyed-in, followed by RETURN. Each line is placed on the screen with the space between the lines being equal. The following example shows the user how to place multiple lines of text and how to change the default values of the text. Refer to Figure 9-8.

1. Select menu item TEXT.
2. Select the type style from the styles shown on the menu.
3. Select menu item HGT to change the height of the text. Key-in the desired height, followed by RETURN.
4. Select menu item SLANT to change the slant of the text. Key-in the angle of the text, followed by RETURN.
5. Select menu item ANGLE to change the angle of the line of text from horizontal. Key-in the angle, followed by RETURN.
6. Select menu item LINE WEIGHT to change the thickness of the line used to form the letters. Key-in the number of the line thickness, followed by RETURN.
7. Select menu item JUSTIFY to change the justification point to left.
8. Select menu item MULT for multiple lines of text.
9. Digitize the point for the paragraph to begin, remembering that the justification point has been moved.
10. Key-in the lines of text, followed by RETURN after each line of text. Figure 8-8 shows multiple lines of text using Roman style type 5 mm high, letters slanted at 15 degrees, lines of text at a 10-degree angle, number 4 line thickness, and justified left.

From these examples it should be apparent to you that there are many different ways that text can be represented with CAD. The speed and neatness that can be gained over lettering by hand should also be apparent. With all the different lettering styles and sizes available to the user it is

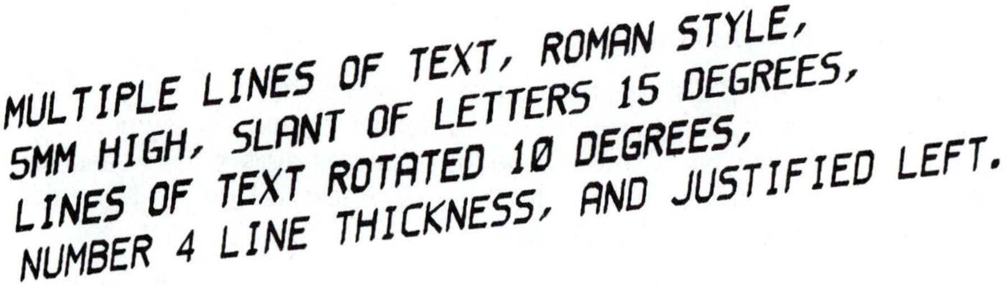

MULTIPLE LINES OF TEXT, ROMAN STYLE,
5MM HIGH, SLANT OF LETTERS 15 DEGREES,
LINES OF TEXT ROTATED 10 DEGREES,
NUMBER 4 LINE THICKNESS, AND JUSTIFIED LEFT.

FIGURE 9-8 Changing text default values.

very easy to become more creative and to add to the neatness and legibility of drawings.

PLACING DIMENSIONS WITH CAD

After a drawing has been produced, it is necessary to place dimensions on the drawing. Dimensions are used to define size and to locate details. Dimensions are one of the most important parts of the drawing because they are used to manufacture or assemble what is drawn. Because there are so many ways in which a dimension may have to be placed on a drawing, most CAD systems have many differing dimensioning styles. Basically, three different types of dimensions are used on drawings: linear, circular, and angular. All these dimensioning styles are covered here.

The placing of dimensions with CAD is a relatively easy task when compared to hand drawing of dimensions. The reason for this is because the computer automates much of the dimensioning process. When drawing by hand it is necessary to draw each extension line and dimension line, all arrowheads, and text. With CAD, dimensioning is a much easier task. The operator digitizes the two points to be dimensioned, digitizes the position for the dimension line, and the computer automatically places the dimension and extension lines, arrowheads and dimension figure. Any drafter having slaved over drawing dimensions by hand cannot help but smile at the ease of placing dimensions on a drawing using CAD.

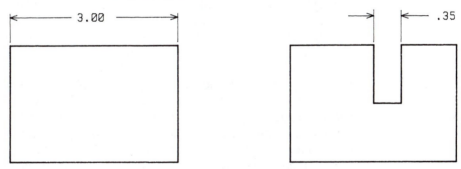

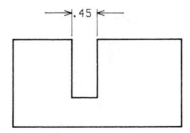

FIGURE 9-9 Typical dimensioning styles.

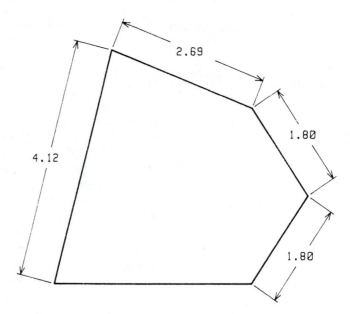

FIGURE 9-10 Dimensions placed at an angle other than horizontal or vertical are drawn horizontally.

Certain default characteristics are associated with dimensioning on most CAD systems. These systems allow the user to set the dimensioning style to be used such as architectural, engineering unidirectional and engineering aligned, and English or metric units. The default values used in this text are engineering aligned, .125 height of text, and decimal inches accurate to two decimal places. Most CAD systems have a predetermined method of placing text, such as the placing of text and arrowheads inside of the extension lines. If there is not enough room for this, the computer will then automatically place the arrowheads outside the extension lines and the text will be inside or the user may have to specify "arrows out" through a menu command. If there is not enough room between the extension lines for the text, then the text is automatically placed outside the extension lines or manually done by the operator. Examples of these features are shown in Figure 9-9. Dimensions placed at some angle other than horizontal or vertical are typically drawn horizontally as shown in Figure 9-10.

Placing Single Linear Dimensions

Linear dimensions can be placed as horizontal, vertical, or parallel to the entities being dimensioned. The menu command used here for single linear dimensions is DIMEN. After this command is selected, the user has only to digitize three points on the drawing to have a dimension placed. Figure 9-11 and the following steps demonstrate linear dimensioning.

1. Select menu item DIMEN.
2. Digitize the two points to be dimensioned.

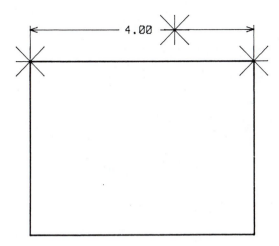

FIGURE 9-11 Placing linear dimensions.

3. Digitize the location for the dimension line. The dimension will then be drawn on screen as shown in the figure. Many systems require the user to specify if the Linear Dimension is horizontal, vertical, or parallel.

User-defined Dimension Values

In placing dimensions, many CAD systems have an extra step that allows the user to input the dimension value instead of having the computer automatically place the dimension. The reason for this is that some CAD systems are extremely accurate when dimensioning. The user may want a linear dimension of 3.000″ for an object but the extreme accuracy of CAD may make the dimension 2.999″. Because of this accuracy, many CAD systems have an extra step that allows users the choice of selecting the computer's measurement or to input their own.

Computers that allow users to input their own values have a set of steps to follow that are different from those shown in Figure 9-11. For inputting user-defined dimension values, refer again to Figure 9-11 and use the following steps.

1. Select menu item DIMEN.
2. Digitize the two points to be dimensioned.
3. Digitize the location for the dimension line.
4. The dimension value will appear in the prompt line allowing the user to change the value or use the computer's value. To select the computer's value the user selects RETURN and the value displayed in the prompt line will be placed on the object. To input another value, the user would key-in the desired value followed by RETURN and the user input value would be displayed.

Placing Multiple Linear Dimensions

An example of placing multiple dimensions on a drawing is shown in Figure 9-12, and explained as follows.

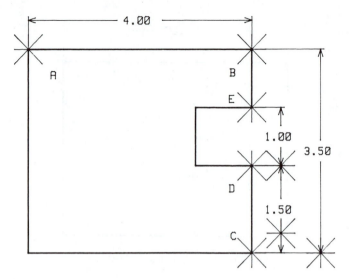

FIGURE 9-12 Placing multiple linear dimensions.

1. Select menu item DIMEN.
2. Digitize corners A and B and the point for the location of the dimension line for horizontal dimension A-B.
3. Digitize points C and D and the point for the location of the dimension line for vertical dimension C-D.
4. Digitize points D and E and the arrowhead of dimension C-D to place dimension D-E aligned with dimension C-D.
5. Digitize points B and C and the location for the dimension for dimension B-C.

Placing Linear Dimensions from a Datum Surface

The datum command is a very convenient function that allows the user to identify a baseline to be used for the placing of dimensions. With this command you are able to reference all dimensions from one baseline or datum line. The menu command used for this function is DIMEN, followed by DATUM. The user then digitizes the baseline, the first dimension point from the baseline, and the position of the dimension line for the first dimension. All other dimensions measured from the baseline are placed on the view by digitizing the next corner from the baseline. The dimension is automatically placed from the baseline at some predetermined distance from the first dimension. Refer to Figure 9-13 and the following steps which describe in detail the use of datum dimensioning.

1. Select menu item DIMEN.
2. Select menu item DATUM.
3. Digitize the baseline (1).
4. Digitize the first corner to be dimensioned (2).
5. Digitize the location for the dimension line (3). The first dimension will be drawn.

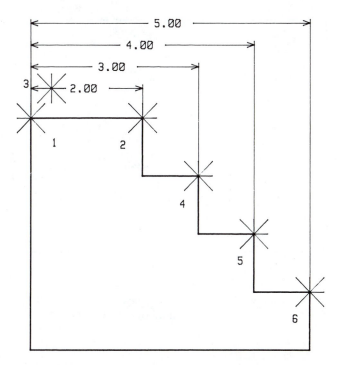

FIGURE 9-13 Placing linear dimensions from a datum surface.

6. Digitize points 4, 5 and 6. As each point is digitized, the dimension is placed on the drawing.

   Other methods of placing linear dimensions are also used, depending upon the type of CAD system in use. However, the methods just described are by far the most common. Other styles include ordinate, isometric, and chain dimensioning. Ordinate dimensioning is measured from a

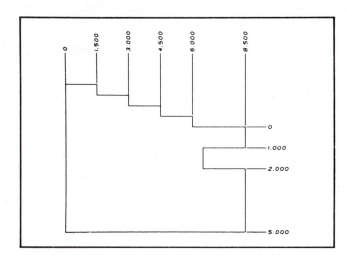

FIGURE 9-14 Ordinate dimensioning. *(Courtesy of Calma Company)*

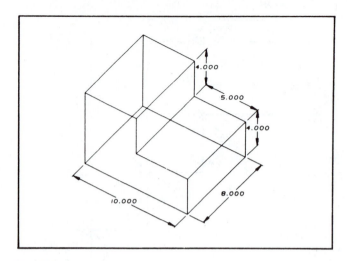

FIGURE 9-15 Isometric dimensioning. *(Courtesy of Calma Company)*

baseline but appears as shown in Figure 9-14. Isometric dimensioning is used to dimension isometric views, as shown in Figure 9-15.

Dimensioning Circles

Circular dimensioning is another class of dimensions used to delineate circles and arcs. Dimensioning circles becomes an easy task using CAD. A circle is dimensioned by using the command DIA. The circle is dimensioned by digitizing the circle then digitizing the position for the dimension text. The computer will automatically draw a line across the diameter of the circle, place arrowheads at the end of the line, and place the text at the digitized point. Some systems will even show the phi (ϕ) sign for the diameter symbol. If the user chooses to use the inside of the circle for the text but there is not enough room, most systems will automatically locate the

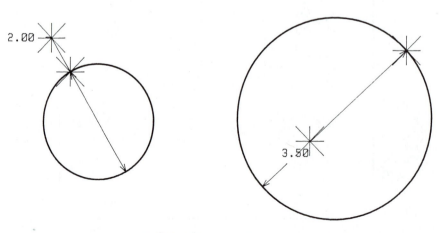

FIGURE 9-16 Dimensioning circles.

text outside the circle. Refer to Figure 9-16 and the following steps to dimension circles.

1. Select menu item DIA.
2. Digitize a point on the circumference of the circle.
3. Digitize the point where you would like the dimension text displayed. The computer will measure, draw, and place the dimension after the second point is digitized. The figure shows how the position of the cursor when digitizing the second point affects the position of the label.

Dimensioning Arcs

The dimensioning of arcs can also be included under the class of circular dimensioning. The placing of radial dimensions follows closely the placing of diametral dimensions. Menu item RADIAL is used here for placing arc dimensions. When the RADIAL command is selected, the user digitizes the arc then the position for the dimension text. The computer automatically calculates the radius, places the pointer and arrowhead, and then adds the text. The second digitized point determines whether the pointer is outside the arc or placed inside the arc from the center of the arc. Figure 9-17 shows internal and external radial dimensioning.

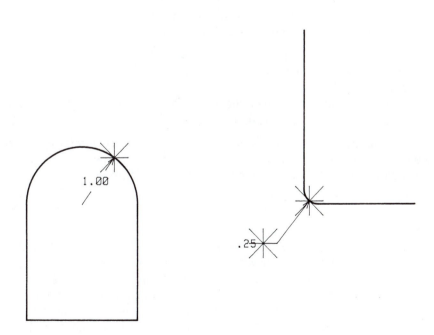

FIGURE 9-17 Placing internal and external radial dimensions.

The following steps describe how to place an external radial dimension.

1. Select menu item RADIAL.
2. Digitize the arc.
3. Digitize the location of the text for the dimension. For the .25″ radius arc, this second digitized point is located outside the arc. This will cause an external radial dimension to be placed on the arc, as shown in the figure.

The following steps show how to place an interior radial dimension.

1. Select menu item RADIAL.
2. Digitize the arc.
3. Digitize the location of the text for the dimension. For the 1″ arc, this point is located between the arc and the arc center. The radial dimension will now be located inside the arc, as shown in the figure.

Placing Angular Dimensions

There are times when it is necessary to dimension the angle between two intersecting lines. This is an easy task when performed by CAD. With most systems, the user individually digitizes the two intersecting lines that make up the angle, then digitizes the location for the text. The computer automatically calculates the angle, and places extension lines, the radial dimension line, and the dimension text. The text can be in either minutes and seconds or decimal degrees. Menu command ANGULAR is used here for dimensioning angles. Figure 9-18 and the following steps demonstrate the placing of angular dimensions.

1. Select menu item ANGULAR.
2. Digitize line A-B.
3. Digitize line B-C.
4. Digitize the location for the dimension text. The computer will place the angular dimension, as shown in Figure 9-18A.

The order in which the two lines are digitized may determine the sweep of the radial dimension line and the angle to be measured. This is demonstrated in the following steps and shown in Figure 9-18B.

1. Select menu item ANGULAR.
2. Digitize line E-F.
3. Digitize line D-E.
4. Digitize the location for the dimension text. The computer will place the angular dimension as shown in the figure. The reason that the large angle is dimensioned instead of the small angle as in the first example is because some CAD systems measure in a counterclockwise direction. In the second example, the line on the left was digitized first causing the computer to measure counterclockwise from that line which makes for the large angle being dimensioned.

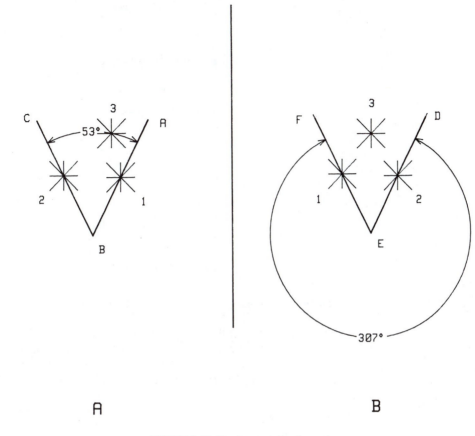

A

B

FIGURE 9-18 Placing angular dimensions.

Changing Default Values for Dimensions

Most CAD systems allow the user to change the height of the text used with the dimensions, the units for measuring, the angular units, the text font (Figure 9-19), and the unit delineator such as mm for millimeters, '' for inches, and so forth. The default values of the dimensions must be changed before the dimensions are placed on the drawing. These values are changed through menu commands before the dimension type is selected. Menu commands DIM HGT and UNITS are used here to make these changes. To change the dimension text height refer to the following steps.

1. Select menu item DIM HGT.
2. Key-in the desired dimension text height, followed by RETURN.
3. Menu item DIMEN, DIA, RADIAL, or ANGULAR is then selected to place the dimensions on the drawing. All the dimension text will now be drawn to the height keyed-in at step 2.

The dimension units can be controlled by using menu function UNITS. Most CAD systems allow the user to choose inches, feet, millimeters,

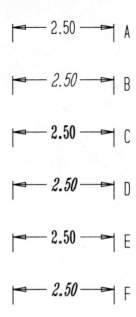

FIGURE 9-19 Different fonts are available on CAD systems for dimension text.

centimeters or meters for the units of measure. After selecting the units, the user must choose the number of decimal places or fractional accuracy. The number of decimal places can be from zero to five, or more, depending upon the CAD system. The fractional accuracy can range from ½ to ¹/₆₄. The user also has the option of displaying the unit delineator with the dimension figure. The steps used to change the units for dimension text are as follows.

1. Select menu item UNITS.
2. The system prompts the user to key-in the units of measure. Change the units by keying-in the units desired, followed by RETURN.
3. The system now prompts the user for the desired decimal places or fractional accuracy. For decimal places, key-in the number desired. For fractions, key-in the fraction desired such as ¹/₃₂.

All dimensions placed after the units have been changed will reflect the units keyed-in after menu item UNITS is selected. Figure 9-20.

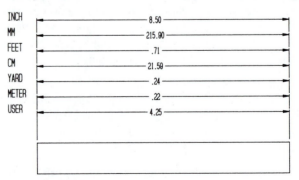

FIGURE 9-20 The same dimension using different units.

Placing Pointers or Leaders

With CAD, circles and arcs are automatically dimensioned with leaders, but there are times when it is useful to use a single line dimension or a leader. For example, when creating an assembly drawing of a part, each part is usually identified with a leader and a letter or number so that the part can be identified from the bill of materials list. By using the single arrow dimension line, the user is able to place leaders on a drawing. Leaders are placed by digitizing the point for the arrowhead then digitizing the end point of the leader line. Menu command LEADER is used here, and Figure 9-21A and the following steps describe how to place a leader line.

1. Select menu item LEADER.
2. Digitize the location for the arrowhead end of the leader.
3. Digitize the end point of the leader line.
4. If desired, digitize another point for a short horizontal connecting line, as shown in the figure, followed by RETURN. The leader line will be placed on the drawing as shown in the figure.

Some CAD systems also have symbols on the menu for adding circles, hexagons, squares or other figures at the end of the leader line. After the leader is placed, the user digitizes the needed symbol on the menu, then digitizes the end point of the leader line. The symbol is drawn onto the end of the leader, and then can be used to contain a number or letter for part identification. An example of a circle drawn onto the end of the leader is shown in Figure 9-21B.

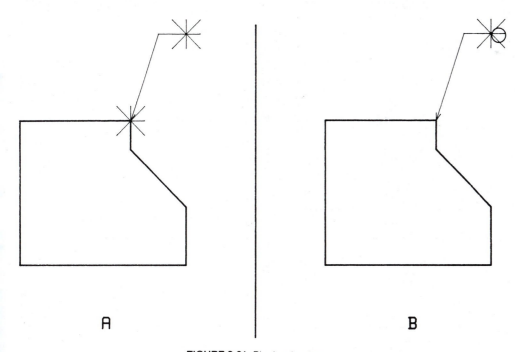

A   B

FIGURE 9-21 Placing leaders.

Placing Dimensions with Tolerances

The design and drawing of parts may require tolerance dimensions to allow for the mass production and interchangeability of parts. A *tolerance* is a range of sizes within which a part can still be functional. Methods of placing tolerance dimensions on a drawing with CAD are common, because of their widespread use in industry. One method used to place tolerance dimensions is by placing them on the drawing as labels separate from the dimension figures. Other systems allow the user to set the upper and lower limits which will automatically be added to the dimension figure when the dimension is placed. Menu command TOLER is used here to demonstrate tolerancing with CAD. The following steps are for automatically placing dimensions with tolerances. Refer to Figure 9-23.

1. Select menu item TOLER.
2. Key-in the values of the upper limit and the lower limit of the desired tolerance, followed by RETURN.
3. Select menu item DIMEN.
4. Digitize the starting dimension point.
5. Digitize the ending dimension point.
6. Digitize the location point for the dimension line. The dimension will be drawn with the dimension figure and the tolerance figures as you input them in step 2.

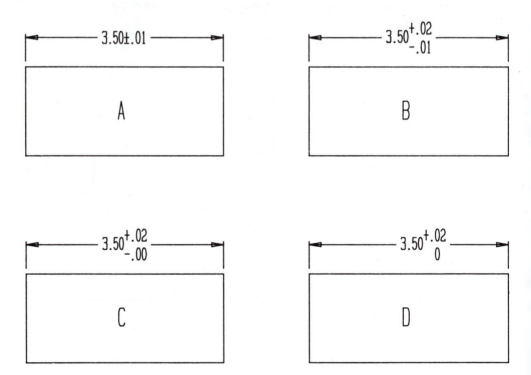

FIGURE 9-22 Different limit tolerance dimensions can be created with CAD.

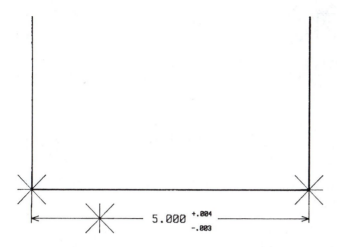

FIGURE 9-23 Placing tolerance figures with dimensions.

Placing Geometric Dimensions and Tolerances

Geometric dimensioning and tolerancing is a system of dimensioning and tolerancing drawings with emphasis on the actual function of the part and its relationship to other parts. This dimensioning technique involves the use of symbols to represent the position or form of tolerance being placed. The feature control symbol is the method used to state the position or form tolerance. Figure 9-24 is a typical feature control symbol used in geometric tolerancing. With CAD, the long rectangular frame might be placed first, followed by the symbol for parallelism. The letter A and the tolerance .0025 would then be placed in the frame by using the TEXT command.

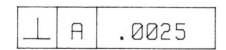

FIGURE 9-24 Typical feature control symbol used in geometric tolerancing.

Most CAD systems provide geometric tolerancing symbols as part of the software package when purchased. If a CAD system does not include these symbols in the software, the user can easily draw them once and add them to the symbols library. If symbols are provided with the software, they are usually part of the menu or can be accessed through a menu function. If they are part of the menu, each symbol is drawn on the menu and the user has only to digitize the symbol on the menu and digitize its position on the drawing. Figure 9-25 shows some of the geometric tolerancing symbols as they may appear on a menu. Figure 9-26 shows a drawing using geometric tolerances.

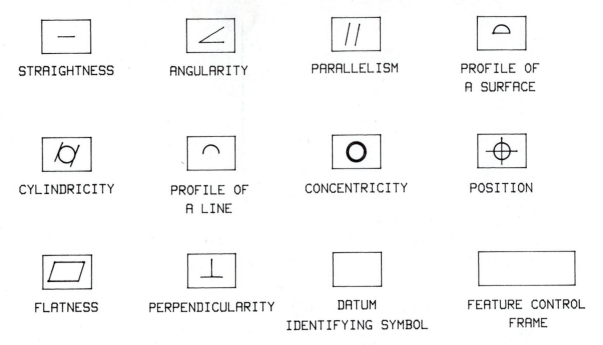

STRAIGHTNESS   ANGULARITY   PARALLELISM   PROFILE OF A SURFACE

CYLINDRICITY   PROFILE OF A LINE   CONCENTRICITY   POSITION

FLATNESS   PERPENDICULARITY   DATUM IDENTIFYING SYMBOL   FEATURE CONTROL FRAME

FIGURE 9-25 Typical geometric tolerancing symbols as they may appear on a menu.

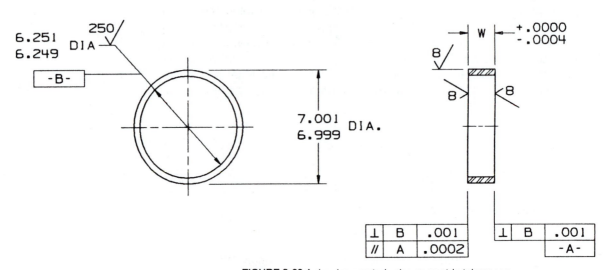

FIGURE 9-26 A drawing created using geometric tolerances.

      The example used in this text has the symbols on the menu. Menu item GEO TOL is used to place these symbols on the drawing. Refer to **Figure 9-27** and the steps that follow to place geometric tolerancing symbols on a drawing.

1. Select menu item GEO TOL.
2. Digitize the position of the datum reference box on the menu.

3. Digitize the position on the drawing for the datum box for datum surface B (1). The user may want to place construction lines or a grid on the screen before placing the box so that the datum box is aligned with the datum surface.
4. Digitize the position of the perpendicularity symbol on the menu.
5. Digitize the position for placing the symbol on the drawing (2).
6. Digitize the position of the datum reference box on the menu.
7. Digitize the corner of the perpendicularity symbol on the drawing to place the datum reference box (3).
8. Digitize the position of the long rectangular box on the menu.
9. Digitize the corner of the datum box placed in step 7 on the drawing to place the long rectangular box used to place the tolerance text (4).
10. Select menu item TEXT.
11. Digitize the position for the letter B in the datum box.
12. Key-in the letter B and RETURN.
13. Digitize the position for the letter B in the upper datum box.
14. Key-in B and RETURN.
15. Digitize the position for the tolerance.
16. Key-in .004 and RETURN to complete the feature control symbol as shown in the figure.

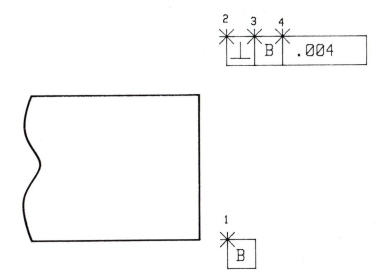

FIGURE 9-27 Placing geometric tolerancing symbols on a drawing.

GEOMETRIC CALCULATIONS
Calculating Distance and Angles

Most CAD systems can determine a number of different geometric calculations of a 2-D drawing. Common features include perimeter, area, centriod, moment of inertia, length of entities, angles of lines, and others. Some CAD systems that can produce true 3-D models can produce other calculations in addition to the ones listed above. However, this chapter will deal only with 2-D drawings, leaving 3-D drawings for Chapter 10.

As an object is being drawn or designed, there are times when the designer must measure a distance or an angle. When drawing on paper this task is performed with a scale or a protractor. With CAD, this would be impossible to do while the drawing is on screen. However, CAD systems have commands that can be used to measure lines and angles. For the purposes of this text, these commands are called MEAS DIST and MEAS ANGLE. The MEAS DIST command can be used to measure a line or part of a line, and the distance between two points or entities. To use this command, the operator digitizes the two points or entities and the computer calculates the distance and displays it in the message line. MEAS ANGLE can be used to measure the angle of a line or the angle between two lines. To use this command, the operator digitizes two points on the line, or one point on each line, and the computer calculates the angle and displays the distance in the message line.

Figure 9-28 and the steps that follow show how to use the MEAS DIST command.

1. Select menu item MEAS DIST.
2. Digitize the end of line A-B.
3. Digitize the other end of the line. The length of the line (3.57) will be calculated and displayed in the message line on the CAD system.

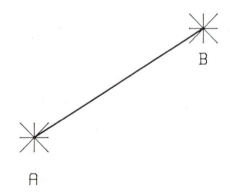

3.57

FIGURE 9-28 Measuring a distance.

Figure 9-29 and the following steps show how to use the MEAS ANGLE command.

1. Select menu item MEAS ANGLE.
2. Digitize a point on line C-D.
3. Digitize another point on line C-D. The computer will calculate the angle (9Ø) and display it in the message line. The order in which the line is digitized is critical in determining the angle. Looking at Figure

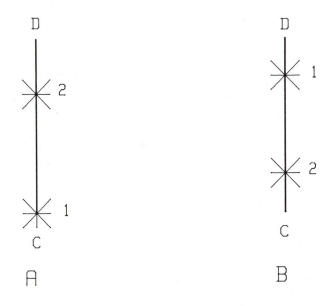

FIGURE 9-29 Measuring the angle of a line.

9-29B, if the first digitized point were above the second, the angle of the line would be measured as 180 degrees. If the first point digitized were below the second point, the angle would be measured as 90 degrees.

Figure 9-30 and the steps that follow show how to use the MEAS ANGLE command to measure the angle between two intersecting lines.

1. Select menu item MEAS ANGLE.
2. Digitize one leg of the angle.
3. Digitize the other leg of the angle. The computer will calculate the angle between the two lines and display the angle (45) in the message line. The order in which the lines are digitized determines the angle measured by the computer. If the bottom line were digitized first, the measured angle would be 45 degrees. However, if the top line were digitized first, the computer would measure the obtuse angle and display 215 degrees. The reason for this is because the computer measures in a counterclockwise direction.

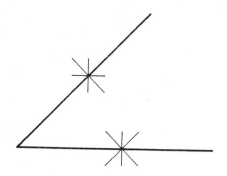

45°

FIGURE 9-30 Measuring an angle between two intersecting lines.

CALCULATING AREAS USING CAD

Calculating areas displayed on screen is another very convenient task that most CAD systems can perform. This is a very useful and time-saving command because it can measure any shape that is bounded on all sides. This command usually has an automatic and manual mode to allow the user the option of manually defining the boundaries of a complicated object.

The menu command used here for this function is CALC AREA. To use this command the user selects the menu option and the computer prompts for automatic or manual area definition. For the automatic command, the user digitizes one of the entities, and the computer calculates the area and displays it in the message line. For the manual command, the user must digitize each entity making up the area, followed by RETURN when the area definition is complete. The computer will then calculate the area and display it in the message line.

Figure 9-31 and the following steps show how the area calculating command is used.

1. Select menu item CALC AREA.
2. Select the automatic routine after the computer prompts you for automatic or manual area definition.

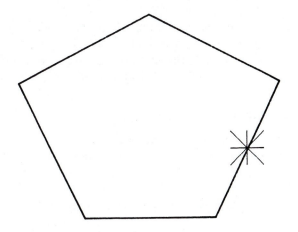

8.000

FIGURE 9-31 Automatically calculating an area.

3. Digitize one of the lines making up the area to be calculated. The computer will calculate the area (8.000) and display it in the message line.

Figure 9-32 and the following steps show how the manual function is used to calculate an area.

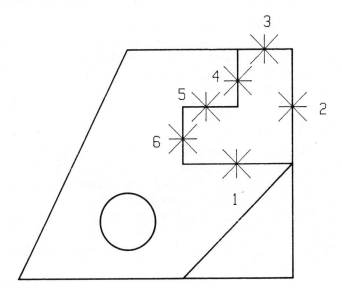

3.000

FIGURE 9-32 Manually calculating an area.

1. Select menu item CALC AREA.
2. Select the manual routine from the prompt line.
3. Digitize in order the entities making up the area to be calculated, followed by RETURN when completed. The computer will then calculate the area (3.000) and display it on the screen.

Calculating Perimeters

The actual or projected lengths of either a single entity or groups of entities can be determined with CAD using a menu item such as PERIM. The perimeter of the part shown in Figure 9-32 can be determined using the PERIM menu option. This is accomplished by digitizing the perimeter entities. The value will then be displayed in the prompt line.

Centroid

The CENTROID option is used to determine the center of gravity of a closed figure. This is usually accomplished following the same basic procedures used to find the area of a closed figure. The menu command that might be used would be CENT or CENTROID. After the entities that surround the closed figure are identified, the centroid's position is marked with an asterisk (*) or some other marker. The X and Y coordinates may also be displayed in the prompt line.

Moment of Inertia

This option is used to determine the moment of inertia of a part. It is used to determine the polar moment of an enclosed area relative to a defined axis. The axis location is defined by digitizing its location and the area is selected by digitizing each entity or by using a window. A prompt will be displayed with the moment of inertia.

Verify Options

Many CAD systems will allow the user to verify or determine various values of an entity such as line type, line weight, pen number, level, color, and so forth. For example, the option can be used to determine the level that an entity has been assigned before deleting all items on a particular level. This option might be assigned the command VERIFY or INQUIRE. To determine the various values assigned to an entity, it must first be identified by digitizing. After the entity is digitized, the properties are listed in the prompt line.

SUMMARY

This chapter covers some very powerful and useful commands available with CAD. These commands make dimensioning and the placing of text much less toilsome for the drafter. With the completion of this chapter, you should be able to draw a simple detail drawing of an object such as those shown in the Drawing Exercises at the end of this chapter. It is important that you attempt to use all of the commands discussed in this chapter for reinforcement and to improve your speed on the CAD system. Chapter Nine shows you some of the special drawing functions that can be used to save additional time and make some drawing functions even easier.

Chapter Nine REVIEW

1. List or sketch some of the hatch symbols used on your CAD system.
2. List the command and the steps used on your CAD system to place a pattern fill.
3. List the command and the steps used on your CAD system to place text.
4. List the commands and the steps used on your CAD system to change the default values of text.
5. List the default values used on your CAD system for text.
6. What is the command used on your CAD system, and what are the steps for placing linear, circular and angular dimensions?
7. In what direction does your CAD system measure when calculating angular dimensions, clockwise or counterclockwise?
8. List the different dimensioning styles used on your CAD system and explain how you can change the dimensioning style.
9. Describe the steps used with your CAD system to place tolerance figures on a drawing.
10. Determine if your system has geometric tolerancing symbols. Describe how such symbols are placed onto a drawing.
11. List the commands on your CAD system to measure distance and measure angle, and describe how they are used.
12. What command is used on your CAD system to calculate area? Describe the steps used for this function.

Chapter Nine DRAWING EXERCISES

1. Using default text values, place your name, affiliation, date, and course on the screen.
2. Change some of the default text values and place a short paragraph of text on the screen.
3. Dimension those drawings selected by your instructor from Chapter Six and calculate the area.
4. From the exercises that follow, draw those selected by your instructor.

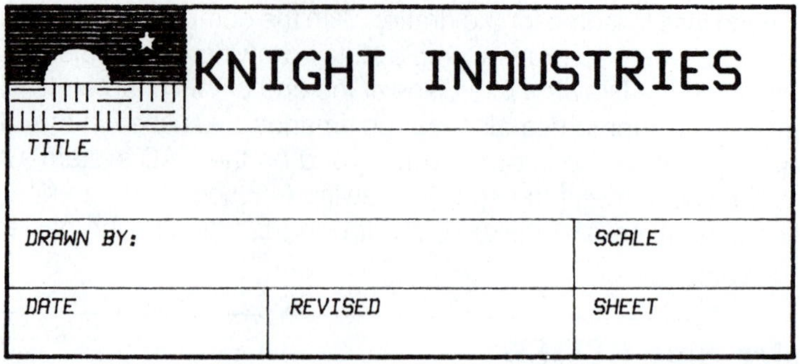

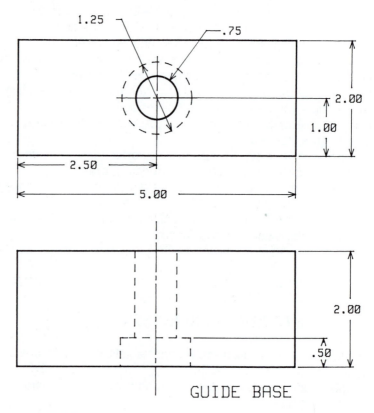

KNIGHT INDUSTRIES

TITLE

DRAWN BY: SCALE

DATE   REVISED   SHEET

FIGURE 9-33 Design a title block, such as that shown, with a logo of your choice, using fill patterns and different text styles.

1.25
.75
2.00
1.00
2.50
5.00

2.00
.50

GUIDE BASE

FIGURE 9-34 Draw the Guide Base, making the front view a full section.

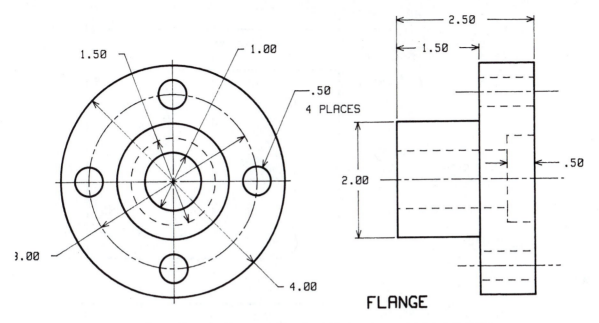

FLANGE

FIGURE 9-35 Draw the Flange, making the right-side view as a full or half section.

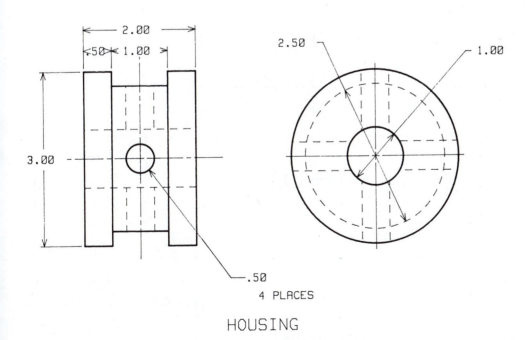

HOUSING

FIGURE 9-36 Draw the Housing, making the left-side view as a half section.

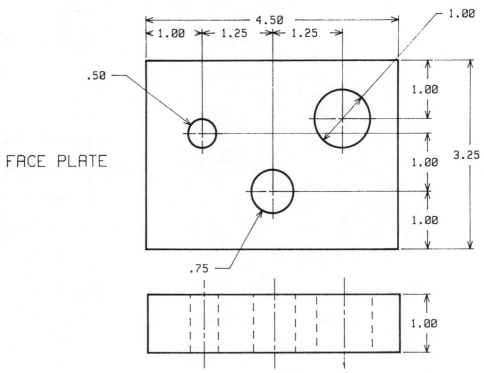

FACE PLATE

FIGURE 9-37 Draw the Face Plate and pass an offset section through the three holes.

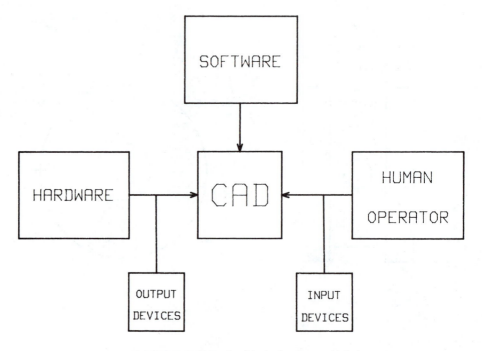

FIGURE 9-38 Draw the block diagram and label.

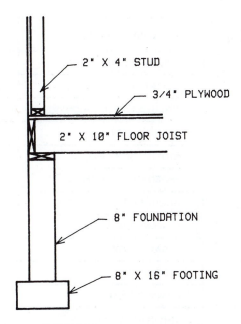

FIGURE 9-39 Draw the wall section.

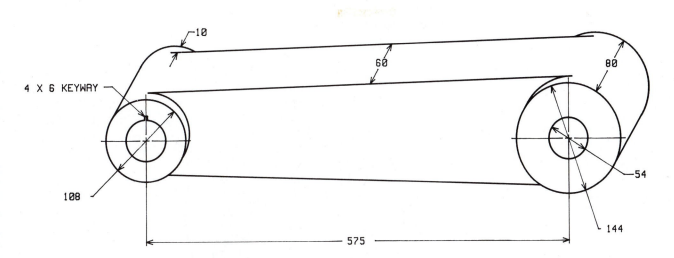

FIGURE 9-40 Draw and dimension the Spanner Arm and scale it to fit on a 297 x 420 sheet (metric).

	TYPE OF TOLERANCE	CHARACTERISTIC	SYMBOL	SEE:
FOR INDIVIDUAL FEATURES	FORM	STRAIGHTNESS	—	6.4.1
		FLATNESS	▱	6.4.2
		CIRCULARITY (ROUNDNESS)	○	6.4.3
		CYLINDRICITY	⌀	6.4.4
FOR INDIVIDUAL OR RELATED FEATURES	PROFILE	PROFILE OF A LINE	⌒	6.5.2 (b)
		PROFILE OF A SURFACE	⌓	6.5.2 (a)
FOR RELATED FEATURES	ORIENTATION	ANGULARITY	∠	6.6.2
		PERPENDICULARITY	⊥	6.6.4
		PARALLELISM	//	6.6.3
	LOCATION	POSITION	⊕	5.2
		CONCENTRICITY	◎	5.11.3
	RUNOUT	CIRCULAR RUNOUT	↗ *	6.7.2.1
		TOTAL RUNOUT	↗↗ *	6.7.2.2
*Arrowhead(s) may be filled in.				

FIGURE 9-41 Create a geometric tolerance symbols library. *(Courtesy of the American Society of Mechanical Engineers, ANSI Y14.5)*

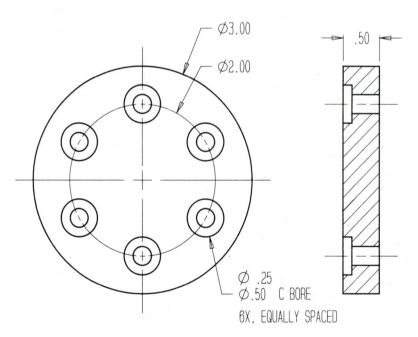

FIGURE 9-42 Draw the Face Plate and section the side view.

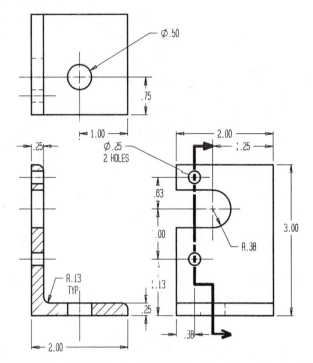

FIGURE 9-43 Make a 3-view drawing of the Angle and create an offset section view.

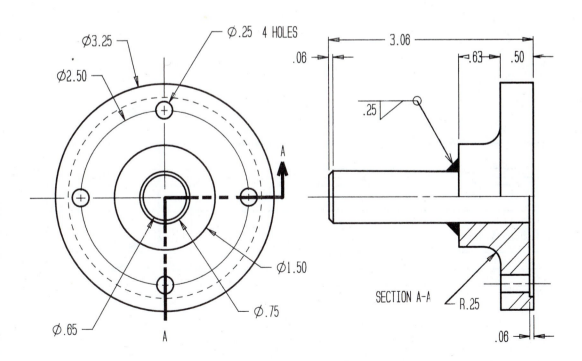

FIGURE 9-44 Draw the Shaft Assembly with the side view as a half section.

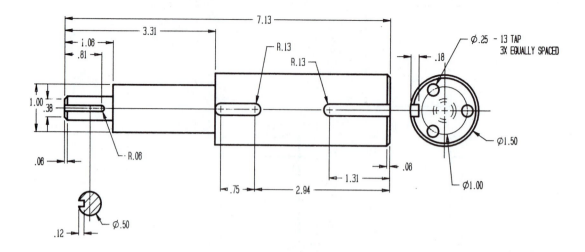

FIGURE 9-45 Draw the Sprocket Shaft and a removed section.

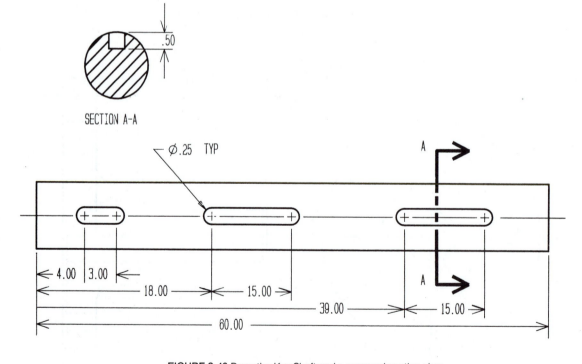

SECTION A-A

FIGURE 9-46 Draw the Key Shaft and a removed section view.

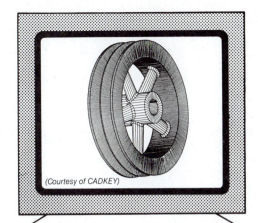

(Courtesy of CADKEY)

Up to this point, you have drawn only simple two-dimensional objects. This chapter describes how to create more sophisticated drawings in a number of different application areas, including three-dimensional (3-D) drawings. The first part of the chapter introduces the reader to the fundamentals of pictorial and 3-D drawings. Applying CAD to mechanical, architectural, electrical, and business graphics is also covered. Some of the more sophisticated and powerful design features that are becoming more common on CAD systems are described. These features include design and engineering functions, such as 3-D solids modeling, mass properties, and design analysis functions. These design functions are described as to their application and how they are used in the design of the product.

PREREQUISITES

Before starting on this unit of instruction, you should be able to:

Use

- the functions of your CAD system, as described in Chapters Five through Nine.

Create

- orthographic views of an object using CAD.

Demonstrate

- a knowledge of axonometric projections.
- a knowledge of Cartesian coordinates using X-Y-Z.

OBJECTIVES

After completing this chapter, you will be able to

- create simple isometric, oblique, one point perspective, and three-dimensional drawings.

SECTION 4

3-D CAD, PRESENT AND FUTURE APPLICATIONS

Chapter Ten

3-D Modeling and Design Analysis

- define solids modeling and describe how it is used in the design of a product.
- define and create a wireframe model.
- define a surface model and describe its creation and use.
- describe some common design analysis functions.

CREATING PICTORIAL DRAWINGS

Up to this point, we have been concerned only with creating two-dimensional (2-D) orthographic drawings. It is often necessary to create drawings or design products using pictorial three-dimensional (3-D) models. This task is simplified if the CAD system has isometric or 3-D software. Even if the system has only 2-D software, pictorial drawings can still be produced. The following section describes how to produce isometric, oblique, and one point perspective drawings with a 2-D drawing package.

Isometric Drawings Using a 2-D System

Isometric drawings are among the most common types of pictorial drawings used in drafting-design. To produce isometric drawings with a 2-D system, one must use polar coordinates, an isometric grid, 30-degree lines, construction lines or some other method to generate isometric lines on screen. The 2″ cube shown in Figure 10-1 was drawn using polar coordinates. The following steps show how to produce this drawing.

1. Select menu item RESET ORIGIN.
2. Digitize the lower center of the screen to move the origin.
3. Select menu item LINE.
4. Select menu item POLAR.
5. Key-in A3Ø,L2 to draw the first isometric line (4-1).
6. Key-in A9Ø,L2 (1-2).

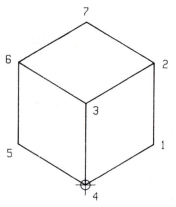

FIGURE 10-1 An isometric drawing can be made with a 2-D CAD system by using polar coordinates.

7. Key-in A210,L2 (2-3).
8. Key-in A270,L2 (3-4).
9. Key-in A150,L2 (4-5).
10. Key-in A90,L2 (5-6).
11. Key-in A30,L2 (6-7).
12. Key-in A-30,L2 (7-2).
13. Select menu item MOVE and digitize corner number 3 on the cube.
14. Key-in A150,L2 (3-6).

Creating Oblique Drawings with a 2-D CAD System

Simple oblique drawings can be produced quickly on a 2-D CAD system by using a grid, polar coordinates or construction lines. The oblique drawing shown in Figure 10-2 was produced using a .50″ grid. The following steps describe how to create the oblique drawing of the figure.

1. Select menu item GRID.
2. Key-in .5 for the grid separation.
3. Select menu item LINE.
4. Digitize the grid points on screen as shown in the figure to produce the oblique drawing.

One Point Perspective Drawings

Simple, one point perspective drawings can be made using a 2-D CAD system. This is accomplished by drawing the front face of the object and then using construction lines to project the depth lines back to a vanish-

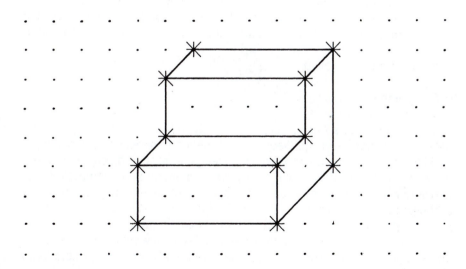

FIGURE 10-2 An oblique drawing can be produced with a 2-D CAD system by using a grid.

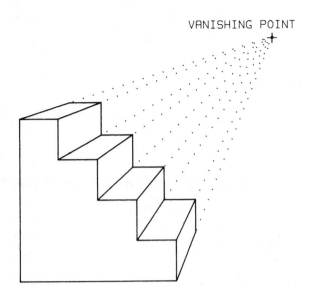

VANISHING POINT

FIGURE 10-3 A one point perspective drawing can be made on a 2-D CAD system by drawing the front face, and then using construction lines to project the depth lines back to a vanishing point.

ing point. Figure 10-3 shows a one point perspective drawing produced by using construction lines projected back to a vanishing point, and drawing over the construction lines for the depth of the drawing.

COMPUTER-AIDED DESIGN

Computer-Aided Design is most commonly associated with 3-D models produced with a CAD system. Various design-related functions can be performed on a 3-D model of a part. These can be grouped into four major areas:

1. Geometric modeling
2. Design analysis
3. Design evaluation
4. Automated drafting

These four areas correspond directly with the last four stages in the traditional design process. Geometric modeling is similar to the concepts and idea stage of the design process. Design analysis corresponds to the analysis and compromise solution. Design evaluation relates to prototypes, and automated drafting corresponds to production or working drawings. The preceding chapters have dealt primarily with the final stage in the design process, production drawings. This is the typical function of CAD and was the primary reason for the development of CAD. However, CAD hardware and software has developed to such an extent that it is now possible to do much more than automated drafting even on microcomputers.

Geometric Modeling

There are a number of different methods to create a model with CAD. Obviously one method is to create a 2-D image as explained in the previous chapters. This 2-D model created with CAD is a *wireframe*. A *wireframe* is a model of a part represented by lines that give the object the appearance of being a model made of wires. See Figure 10-4. Some CAD systems have an automatic hidden line removal feature that will search the geometric data base of the part and either remove hidden lines or change them to dashed lines. Other systems require the user to selectively erase hidden lines using the DELETE or ERASE command. There are three types of wireframe models commonly found on CAD systems: 2-D, 2½-D, and 3-D.

3-D Wireframe Models

3-D wireframe modeling is common even among microcomputers. Designing in 3-D is a natural process that may have to be relearned by people experienced in designing using orthographic projection. After the 3-D model of the part is created, the orthographic views can be produced by extracting the necessary views by transforming or manipulating the model.

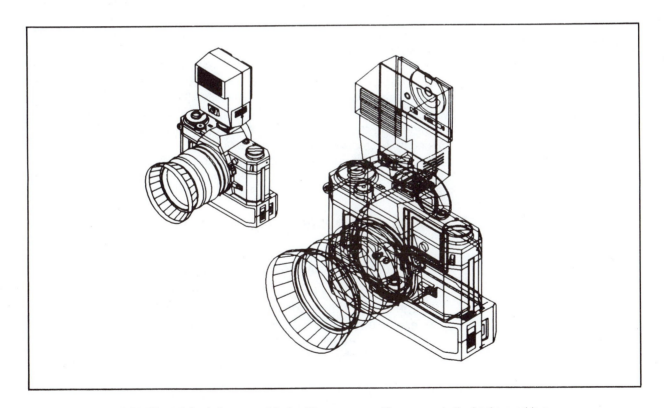

FIGURE 10-4 A wireframe model of a 35-mm camera. The camera in the background is a wireframe with hidden lines removed. *(Courtesy of Calma Company)*

The steps used to create a drawing using a 3-D package vary among different CAD systems. One common method is to draw a wireframe model using X-Y-Z coordinate values. This can be done through extrusion, Point-to-Point, Rotation, or Primitive shape. The object appears to be transparent with no hidden lines. See Figure 10-4. This 3-D model can then be rotated about any axis to be viewed in any position. The orthographic views could then be produced by transforming or manipulating the model so the view point is perpendicular to the face of the object.

To avoid making orthographic views of an object separately from the wireframe model, some CAD systems have the capability of producing the orthographic drawing at the same time that the 3-D model is being created. This is done by dividing the screen into different areas, and assigning orthographic views and the 3-D view to particular areas of the screen. Another method used to produce a 3-D model with geometric modeling is to build the part using primitive shapes. All of these methods will be described further.

As reviewed in Chapter Five, the three axes of the Cartesian coordinates used to produce a 3-D model are as shown in Figure 10-5. The X axis is horizontal, the Y axis is vertical and the Z axis is perpendicular to the screen.

In order to view a model in 3-D, it is necessary to rotate these axes on the 2-D screen. The operator does this by entering the angles of rotation about each of the axes. Positive angles cause counterclockwise rotation, and negative angles cause clockwise rotation. For example, to produce an isometric drawing, the operator enters a value of −45 degrees for the Y axis and 35 degrees 16 minutes for the X axis. This causes the axes to become rotated as shown in Figure 10-6. The user then keys-in X,Y,Z coordinate values for the end points of lines making up the object. The points in the example and their respective coordinate values are shown in Figure 10-7. Once the 3-D model is produced, the operator can view it from any direction by changing the angles of rotation for each of the axes.

Creating the 3-D Model — Point-to-Point

The following steps and Figure 10-8 show how a 3-D model can be produced using Cartesian coordinates and the Point-to-Point method.

1. Select the menu function for 3-DIMENSIONS.
2. The operator must choose the angles of rotation for each axis. In this example, an isometric drawing is produced by keying-in AX35 and AY−45 to revolve the axes. Most systems will display the X,Y,Z axes on screen as shown at the lower left in the figure.

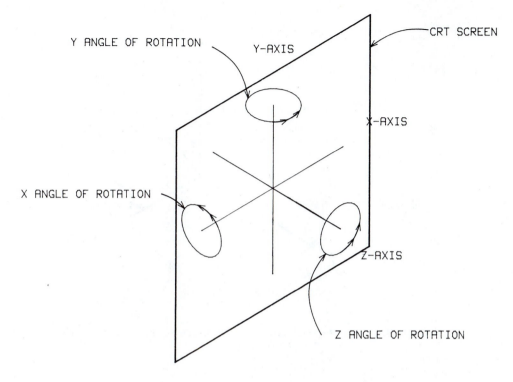

COUNTERCLOCKWISE ANGLE IS POSITIVE

CLOCKWISE ANGLE IS NEGATIVE

FIGURE 10-5 The position of the three axes before rotation, and the direction in which an axis is rotated for 3-D drawings.

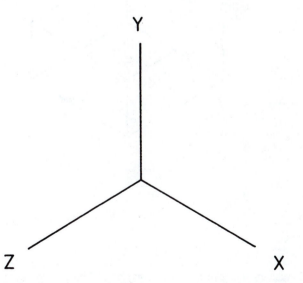

FIGURE 10-6 Position of axes after rotation for a 3-D isometric drawing.

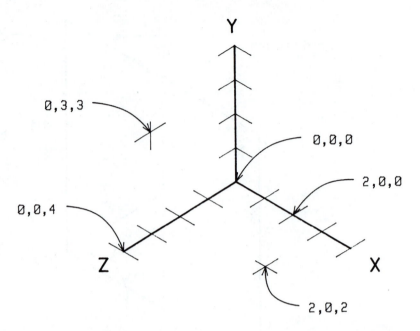

FIGURE 10-7 The three axes showing the coordinate values assigned to different points.

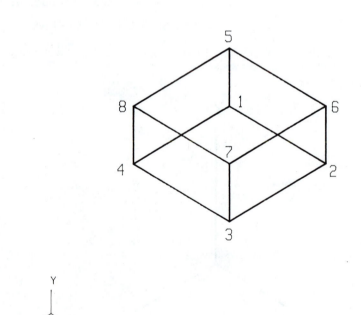

FIGURE 10-8 3-D wireframe model created by keying-in X,Y,Z coordinate values for end points of lines.

The following table shows the coordinate values used to draw the object in Figure 10-8. The table identifies the end point of the lines, the X,Y,Z values for that point, and whether it is a MOVE command or a LINE drawing command.

POINT	X	Y	Z	COMMAND
1	Ø	Ø	Ø	MOVE
2	2	Ø	Ø	LINE
3	2	Ø	2	LINE
4	Ø	Ø	2	LINE
1	Ø	Ø	Ø	LINE
5	Ø	1	Ø	LINE
6	2	1	Ø	LINE
2	2	Ø	Ø	LINE
6	2	1	Ø	MOVE
7	2	1	2	LINE
3	2	Ø	2	LINE
7	2	1	2	MOVE
8	Ø	1	2	LINE
4	Ø	Ø	2	LINE
8	Ø	1	2	MOVE
5	Ø	1	Ø	LINE

After the wireframe has been created, it is saved so that the original model is not changed when doing design modifications. This wireframe can be rotated if the designer feels that the model should be in a different position. Once the final position is chosen, the operator must remove hidden lines. One of the problems inherent in wireframe models is that of reverse imaging. After working on a complex shape, such as the one shown in Figure 10-4, the operator may find it difficult to recognize which lines make up the front of the object and which lines make up the back. Even the simple object shown in Figure 10-8 can be confusing. For clarification, ask yourself which line is closer to the viewer, 1-5 or 3-7? Line 3-7 is chosen as closest to the viewer, leaving lines 1-2, 1-5, and 1-4 as hidden. These three lines are then removed, leaving a wireframe model as shown in Figure 10-9.

Creating the 3-D Model — Extrusion

This is one of the most common, and possibly the easiest, of the methods of creating a 3-D wireframe model of a simple part. It is accomplished by copying a surface and placing it at a new depth. Corners of the surface are joined with lines to create a 3-D model of the part. Figures 10-10 through 10-12 and the following steps show how to use the extrusion method to produce a wireframe model of a part.

1. One face of the object is created using the LINE, CIRCLE, and FILLET commands.
2. The view-point is changed to see the object in isometric as shown in Figure 10-11.

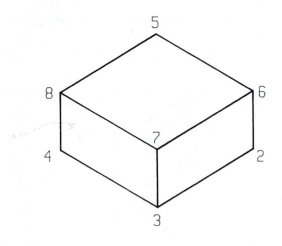

FIGURE 10-9 Model after hidden lines have been removed. The surfaces bounded by the wireframe can now be shaded or colored to produce a solids model.

3. The object is then extruded by using the COPY command and assigning the copy to a new depth. The original drawing was produced at a Z-depth of zero. Place a window around the view to identify the entities to be copied.
4. Input the new Z-depth that the object is to be copied to. For this example input 1 and press RETURN. The 3-D model of the part will be extruded as shown in Figure 10-12.

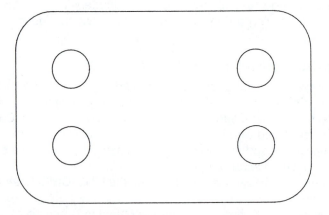

FIGURE 10-10 Orthographic view of the Rest Plate.

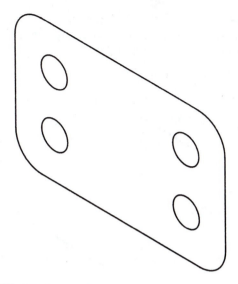

FIGURE 10-11 The Rest Plate is viewed in isometric before extrusion.

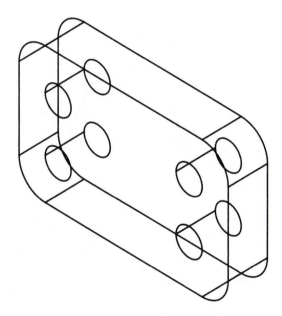

FIGURE 10-12 The Rest Plate is copied and joined causing the 2-D drawing to be extruded into 3-D model of the part.

Extrusion with Scale

Extrusion can be combined with a new scale to produce cones, prisms, and pyramids. The extruded copy of a rectangle can be scaled to half size and placed at a new depth to create a pyramid. Figure 10-13 shows the base of the pyramid before extrusion. Figure 10-14 shows the pyramid after the base has been copied, scaled to one-quarter size, and extruded a distance of 4. Extrusion can take place along any of the three axes. For this example, the part was extruded along the Y axis.

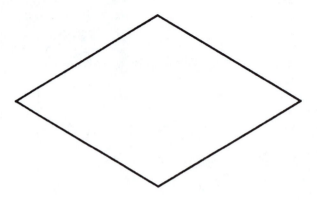

FIGURE 10-13 An isometric view of a base to be scaled, copied, and joined into a 3-D model.

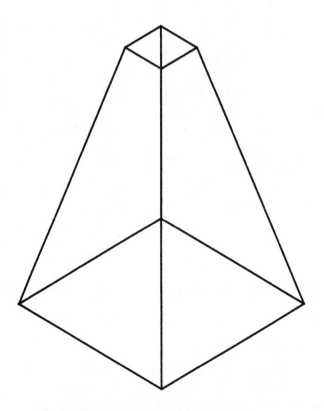

FIGURE 10-14 A pyramid created from the base.

Creating the 3-D Model — Rotation

The rotation method of creating a 3-D wireframe is especially useful when designing symmetrical parts such as a torus, flat and V-belt pulleys, gears, containers, wheels, sheaves, and so forth. A cross section of the part to be produced is first drawn as shown in Figure 10-15. This cross section is then rotated about the center line. The model is produced by copying the profile, specifying the number of copies to make, and the degree increment that the copies are to be rotated. To produce the torus shown in Figure 10-16, eighteen copies were rotated every 20 degrees.

FIGURE 10-15 the cross section of the torus and the location of the center line.

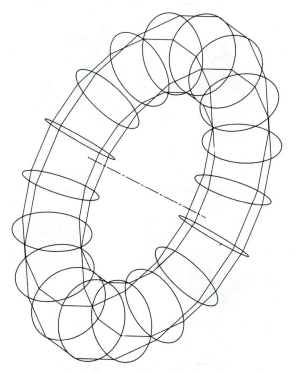

FIGURE 10-16 An isometric view of the torus created by rotating the cross section about the center line.

Creating the 3-D Model — Primitives

The method of using primitive geometric shapes to produce a model can also be used. Figure 10-17 shows some common primitive shapes that can be used to create models. Primitive shapes are placed on a drawing by selecting the primitive from the menu, entering the length, radii, rotation, or other parameter, and digitizing or inputting the absolute or relative coordinates. These primitive shapes can be scaled, rotated, and combined with other primitives or 3-D models to create very sophisticated parts. See Figure 10-18.

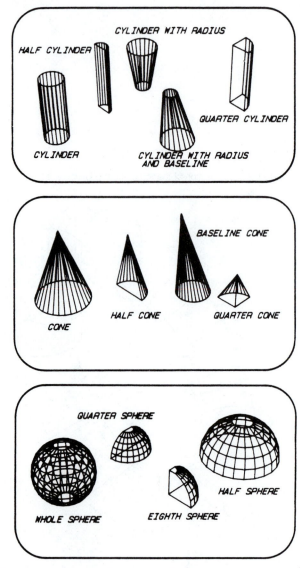

FIGURE 10-17 Common primitive wireframe shapes used to create a 3-D model. *(Courtesy of VersaCAD)*

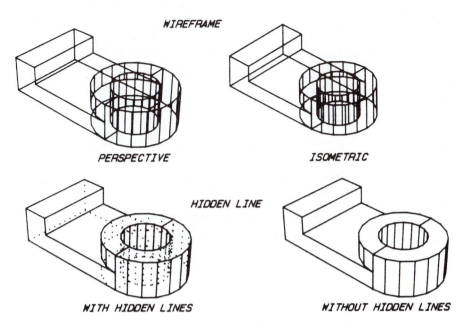

WIREFRAME

PERSPECTIVE   ISOMETRIC

HIDDEN LINE

WITH HIDDEN LINES   WITHOUT HIDDEN LINES

FIGURE 10-18 Wireframe model viewed at different angles and shown with and without hidden lines. *(Courtesy of VersaCAD)*

Editing the 3-D Wireframe Model

After the model is created, it may have to be edited to remove hidden lines, change the viewing angle, or to produce orthographic views. Common editing commands include MOVE, COPY, ROTATE, SCALE, DELETE, and HIDE. After the 3-D model has been created, the orthographic views may have to be produced. This can be done by viewing the object perpendicular to its faces. The 3-D model displayed in Figure 10-12 will be manipulated to create the orthographic views. This is done by rotating the object or selecting an assigned view from the menu. Some CAD systems will automatically generate the six orthographic views of a part. Selecting FRONT will automatically display the front view of the part as shown in Figure 10-19. This view would then be saved as a pattern or symbol. The top and right side view would then have to be created and saved.

After the orthographic views have been created, they must be retrieved and placed on a drawing sheet as shown in Figure 10-20. Further editing of the drawing, such as the addition of hidden lines, dimensions, notes, and labels may be necessary.

Creating Orthographic and 3-D Drawings Simultaneously

Another method used to create a 3-D model is to make it at the same time that the orthographic views are being drawn. Using the third angle of projection, the six views of an object are arranged as shown in

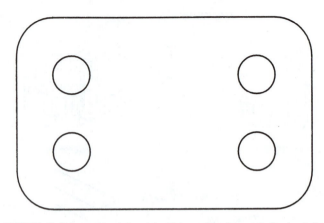

FIGURE 10-19 Isometric view rotated into position to create the front view.

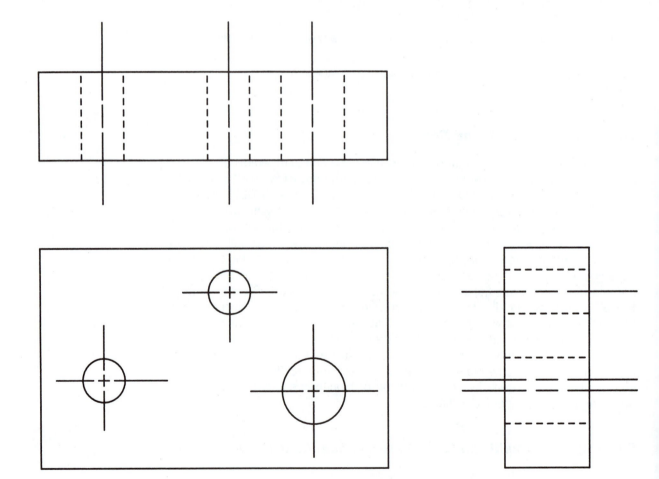

FIGURE 10-20 Edited 3-view orthographic drawing created from the 3-D wireframe model.

Figure 10-21. The operator has the option of choosing which of these six orthographic views to display on screen when producing the 3-D model. This is done by dividing the screen into the number of views desired, including the 3-D model. Figure 10-22 shows a screen divided into four parts, with labels showing the view to be produced in each section. The operator may assign a view to any section of the screen. For example, the front view in the figure could have been displayed in the upper right corner if the user had so chosen.

After the views and screen areas are chosen, the user must enter the angles of rotation for the 3-D model. The operator then draws the model using X,Y,Z coordinates. Each entity is added to each view as it is entered. The model produced is a wire model because all of the lines are drawn solid. If desired, the user must edit the model after it is complete to show hidden features. Figures 10-23 through 10-27 and the following steps demonstrate how a 3-D wireframe model and orthographic views are made.

1. The user inputs the number of views to be displayed (4).
2. The operator then chooses the location of the views on screen.
3. The angles of rotation for the axes are then entered (X35,Y−45).

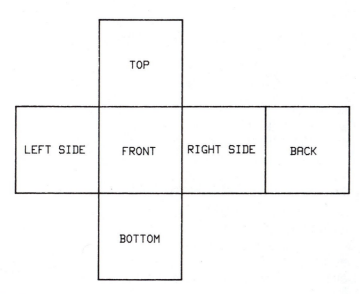

FIGURE 10-21 Six views used in the third angle of projection. The operator can choose which views are to be displayed on screen.

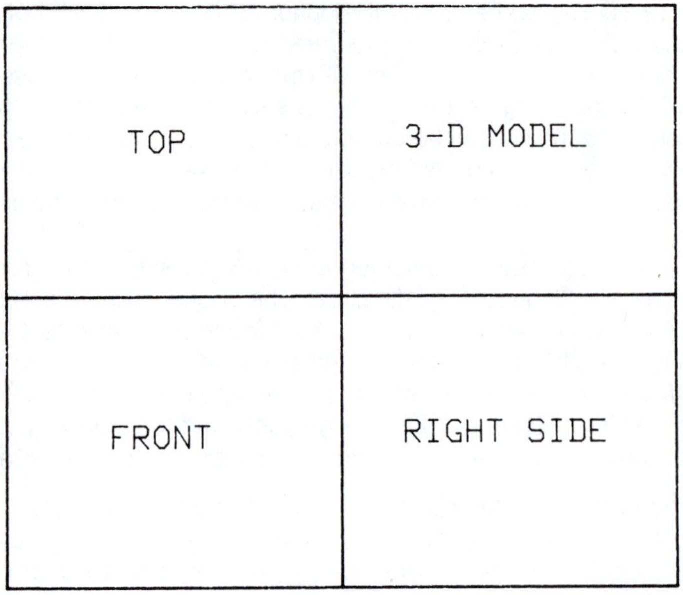

```
+------------------+------------------+
|                  |                  |
|                  |                  |
|      TOP         |    3-D MODEL     |
|                  |                  |
|                  |                  |
+------------------+------------------+
|                  |                  |
|                  |                  |
|     FRONT        |   RIGHT SIDE     |
|                  |                  |
|                  |                  |
+------------------+------------------+
```

FIGURE 10-22 Example of how the screen might be divided to produce orthographic views and a 3-D model.

DRAWING THE FRONT SURFACE, Figure 10-23

4. Select menu item LINE.
5. Key-in XØ,YØ,ZØ.
6. Key-in X2,YØ,ZØ.
7. Key-in X2,Y1,ZØ.
8. Key-in XØ,Y1,ZØ.
9. Key-in XØ,YØ,ZØ. This completes the front surface.

DRAWING THE TOP SURFACE, Figure 10-24

1. Select menu item MOVE.
2. Key-in XØ,Y1,ZØ.
3. Select menu item LINE Key-in XØ,Y1,Z−1.
4. Key-in X2,Y1,Z−1.
5. Key-in X2,Y1,ZØ. This completes the top surface.

DRAWING THE RIGHT-SIDE SURFACE, Figure 10-25

1. Select menu item MOVE.
2. Key-in X4,YØ,ZØ.
3. Select menu item LINE. Key-in X4,YØ,Z−1.
4. Key-in X4,Y1,Z−1. This completes the right-side surface.

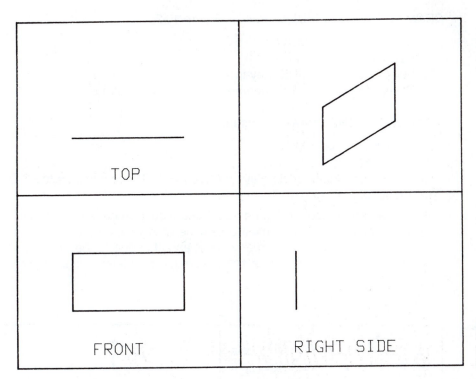

FIGURE 10-23 Drawing the front surface of the box and displaying the orthographic views of that surface.

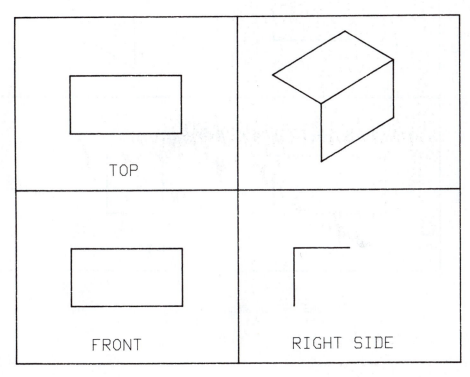

FIGURE 10-24 Drawing the top surface of the box.

DRAWING THE LEFT-SIDE SURFACE, Figure 10-26

1. Select menu item MOVE.
2. Key-in X0,Y0,Z0.
3. Select menu item LINE. Key-in X0,Y0,Z − 1.
4. Key-in X0,Y1,Z − 1. This completes the left-side surface.

COMPLETING THE DRAWING, Figure 10-27

1. Select menu item MOVE.
2. Key-in X0,Y0,Z−1.
3. Select menu item LINE. Key-in X4,Y0,Z − 1. This completes the 3-D model of the box.

     As can be seen in Figure 10-27, the model produced has no hidden lines. Thus, the operator must determine which lines are hidden. This model can now be viewed from any angle and changed to the designer's specifications. How this model is analyzed is covered later in this chapter.

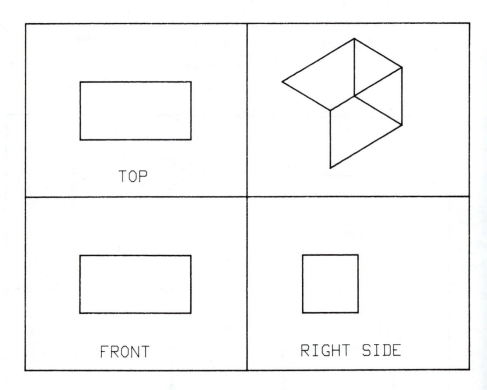

FIGURE 10-25 Drawing the right-side surface.

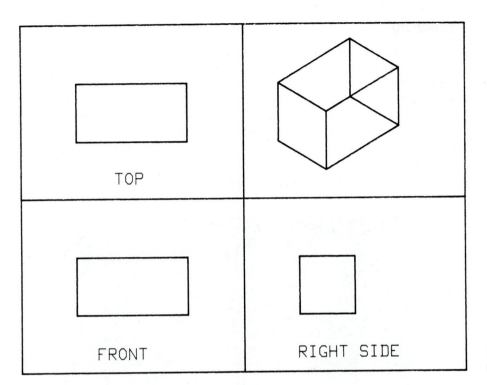

FIGURE 10-26 Drawing the left-side surface.

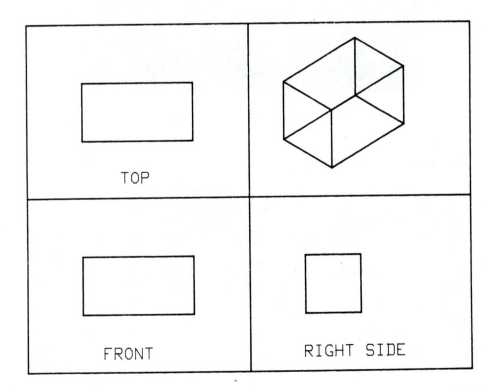

FIGURE 10-27 Completing the 3-D model of the box.

Summary

Most CAD systems use some form of 3-D wireframe modeling. 3-D modeling is a great improvement over 2-D CAD systems. For curved or warped surfaces, contour lines or ruled surfaces can be created. This is especially useful in the design of aircraft and automobile bodies. See Figure 10-28. There are limitations to wireframe models. This becomes apparent if the user desires to analyze the model to determine mass properties and other characteristics of the part. Wireframe models have limited use in manufacturing and for very complex surfaces. These limitations are the strengths of surface and solid modeling.

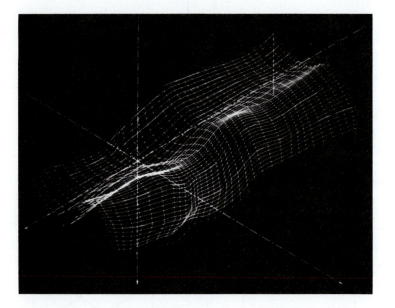

FIGURE 10-28 Complex wire-frame surface model. *(Courtesy of Applicon)*

SURFACE MODELING

Surface modeling is a graphical technique used to define and describe surfaces. A wireframe model can only describe the edges or boundaries of a part. Points between the boundaries of a part cannot be defined with wireframes. Surface models define not only the edges of a part, but also the surface between the edges.

Surface models are produced after the wireframe boundaries have been created. The surfaces between the wireframe boundaries are then defined. After the surface is defined it can be displayed with or without hidden lines. Each surface can then be shaded in different tones of gray or colors. The color photographs insert shows examples of color-shaded models produced with several different CAD systems. The ability to create surface models is becoming popular on microcomputer-based systems. Figure

10-29 shows a shaded surface model of a part.

      After a model has been surface shaded it is possible to change the light source so that the shading will change. There are also smoothing techniques that can be used to eliminate sharp boundaries and create a more realistic image.

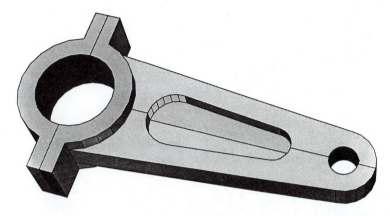

FIGURE 10-29 Shaded surface model of a part with the light source located above and to the right. *(Courtesy of VersaCAD)*

Producing a Surface Model

      To produce a surface model, the wireframe model of the part must be created. This could be done using any of the methods described under wireframe modeling. Figure 10-30 is a wire-frame model that will be used as an example for surface shading. Typical menu commands used for surface modeling include SHADE, ILLUMINATE, LIGHT SOURCE, and INTENSITY.

      The first step in shading a surface would be to set the INTENSITY of the light source. The SOURCE of the light would then be specified using X, Y, and Z coordinate values. The SHADE command would then be used to display the shading using the specified intensity and source of light. See Figure 10-31.

SOLID MODELING

      A solid model looks similar to a surface model but it offers many advantages. It is a mathematically complete representation of a part. Solid modeling requires a great deal of computer power in terms of speed and memory. Improvements in hardware and software will lead to its introduction on microcomputers. However, most solid modeling is performed on mini or mainframe computers. A solid model differs from a surface model by revealing information about the interior of the part. As the model is created, different properties such as mass and material can be assigned to the model. There are two basic approaches to solid modeling: constructive solid geometry (CSG) and boundary representation (B-rep.).

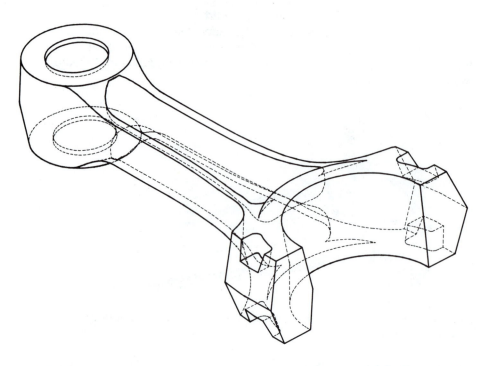

FIGURE 10-30 Wireframe model of a crankshaft before the hidden lines were removed. *(Courtesy of Matra Datavision)*

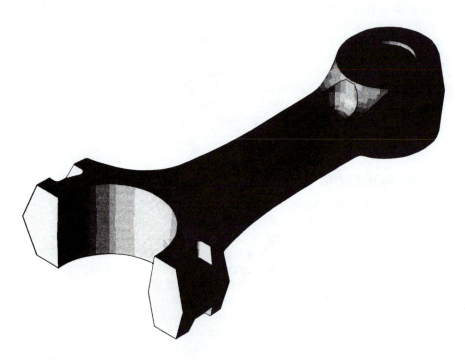

FIGURE 10-31 Shaded surface model of the crankshaft. *(Courtesy of Matra Datavision)*

Solid Modeling — B-rep

Boundary representations are a collection of surfaces, points, and curves. These entities are then combined to create a solid image. B-rep models can be used for complex and unusual shapes such as aircraft wings and fuselages.

Constructive Solid Geometry — Primitive Solids

CSG modeling is a process used to design complex shapes using such simple solid geometric primitives as blocks, cylinders, cones, spheres, and other shapes as shown in Figure 10-32. Primitives are combined through the use of Boolean operations of union, intersection, and difference. CSG works well for regular geometric shapes that can be easily created using primitive solids.

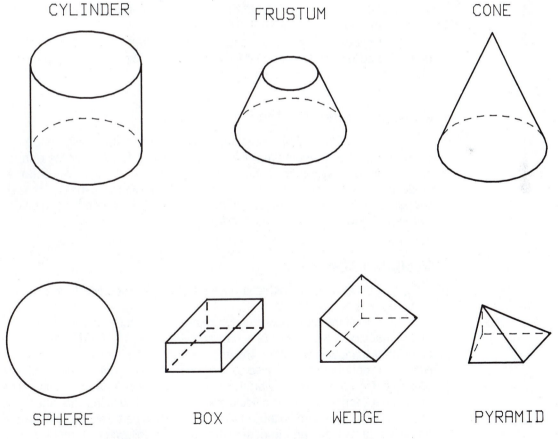

CYLINDER   FRUSTUM   CONE

SPHERE   BOX   WEDGE   PYRAMID

FIGURE 10-32 Typical primitive shapes used with a solids modeling package.

The size of the primitive is defined by inputting the dimensions of the desired solid. For a partial solid, the difference in the dimensions of the shapes is used. For example, to drill a hole in a rectangular block, the user uses the volume of the box minus the specified volume of the cylindrical hole, as shown in Figure 10-33.

After the model is produced, drawings can be automatically created from any viewpoint. Sectional views also can be automatically generated from the model. Design analysis functions can be performed from the model's geometric data base. Some modeling packages provide full color shaded pictures of the model from any viewpoint. The user can also control the light source to produce shading on each surface. This results in virtually photographic quality models which can be displayed on high-resolution color terminals.

Summary

One disadvantage of modeling is the tremendous amount of computer memory needed to create a part. This problem will become less severe as the price of computer-memory drops and hardware becomes more sophisticated. As this occurs, the use of 3-D drawings and modeling will become the norm rather than the exception.

Because there are advantages to the B-rep and CSG methods of producing solid models, hybrid systems have been developed which can combine both modeling techniques.

Computer-aided Design

With a CAD/CAM system, a drafter/designer starts with an idea or a rough sketch and begins making the geometric data base for the part on the graphic display. Often, such a drawing is made as either a wireframe model or a solids model. When the designer is satisfied with the 3-D model, the part can then be run through various design analysis programs to calculate its behavior under various conditions.

Design Analysis

Design analysis includes analysis of mass properties and finite element analysis of 3-D models.

One of the most powerful design analysis features is the *FINITE ELEMENT* method. A 3-D model of a part is divided into a large number of rectangular or triangular elements. The intersections of these elements are called *NODES*. Properties are assigned to the model and the entire object can be analyzed for stress-strain, vibration, heat, and other characteristics. The behavior of each node can be calculated and combined with the characteristics of all the nodes to determine the behavior of the whole part. This information can be shown graphically and is demonstrated in the following example. A load (stress) is imposed on the model to determine the amount of deformation

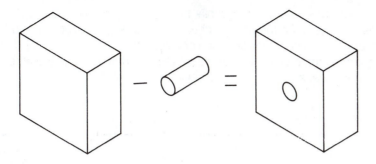

FIGURE 10-33 One method of using solids modeling to show a hole in a box is to subtract a cylinder of specified volume from the box.

(strain). The amount of deformation can be displayed on screen, as shown in Figure 10-34. Some systems have the ability to display dynamically the effects that a load has on a part. As the load is applied, the changes in deformation of the model are displayed and continually updated on screen. The stress placed on the part can also be displayed on screen in the form of a contour plot, as shown in Figure 10-34. A contour plot of stress can be displayed in color, to represent different levels of stress. This process allows the user to determine quickly the highest levels of stress, which are the areas most likely to fail.

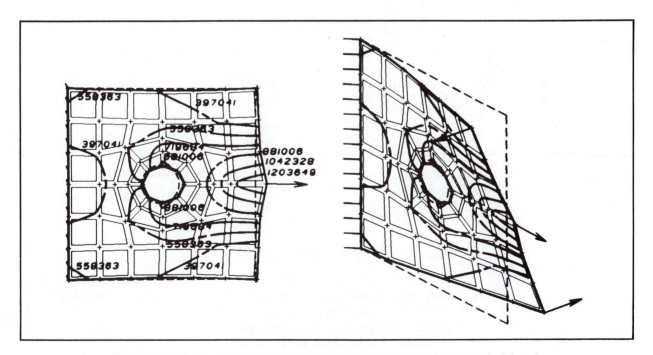

FIGURE 10-34 Finite element model of a plate showing the stress contour plot on the left, and the deformed plate on the right. *(Courtesy of Calma Company)*

Alternate designs are easily produced by changing the 3-D model. Because it is easy for the user to check the model, prototypes are no longer necessary, because the designer can check the design with the computer. The designer can then check the fit of the part for interference with the other parts before the actual assembly of the product, Figure 10-35.

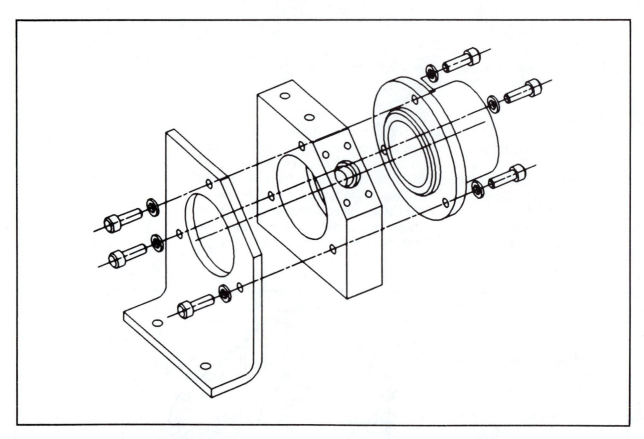

FIGURE 10-35 Submodel parts can be combined into the whole assembly as shown in this exploded view. *(Courtesy of Calma Company)*

Mass Properties Calculations

Once the solids model is produced, various mass properties can be calculated. These calculations are based on the volume and mass densities assigned to the model by the user. Some of the mass properties that can be determined from a solids model are the volume, weight, center of gravity, moments of inertia, area, perimeter, and inertia.

Kinematics

Some CAD systems can display the behavior of moving parts. Kinematic software can animate the motion of simple mechanisms. This capability allows the designer to check for interference of parts. See Figure 10-36.

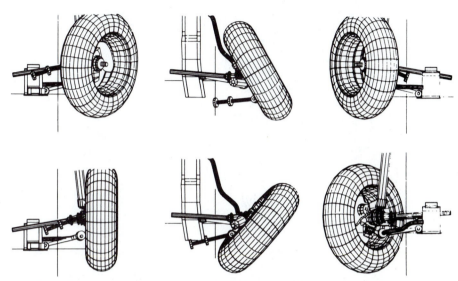

FIGURE 10-36 Interference checking and verification of a tire and suspension system. *(Courtesy of Matra Datavision)*

Automated Drafting

After the part has been designed, engineering drawings are needed for documentation. To produce orthographic drawings, the user moves the point of site and the views are generated from the 3-D model. The operator must then clean up the views produced by the computer, because of inherent difficulties associated with computer-generated views. Dimensions and text are added to the views. Isometric views can also be produced by changing the point of site on the model. Cutaways and exploded views can be generated from the 3-D model. Figure 10-37 shows an isometric view of a solids model, and Figure 10-38 shows a cutaway view of a solids model. Plots can now be made of the completed drawings.

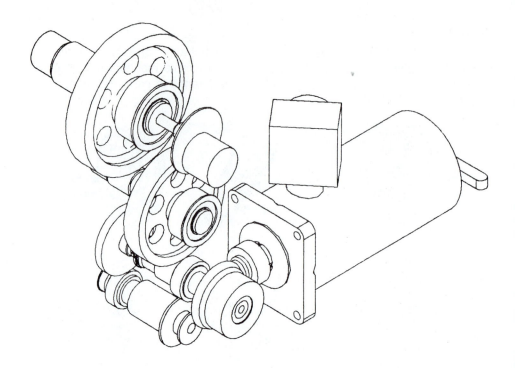

FIGURE 10-37 Isometric view of a solids model. *(Courtesy of MAGI SynthaVision® Solids Modeling System)*

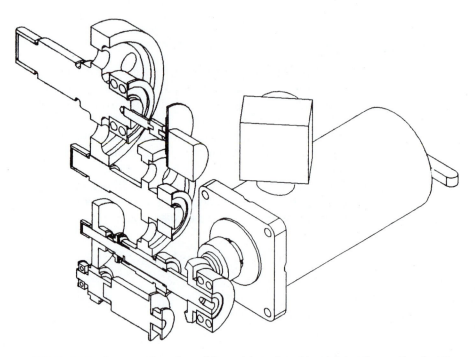

FIGURE 10-38 Cutaway view of a solids model produced by defining the position that the cutting plane will pass. *(Courtesy of MAGI SynthaVision® Solids Modeling System)*

Chapter Ten GLOSSARY

Solids modeling—a process for designing using combinations of solid geometric primitives, such as blocks, cylinders, cones and spheres.

Surface Modeling—A graphical modeling technique used to define and describe surfaces.

Wireframe—a 3-D model of a part represented by lines giving the object the appearance of being a model made of wires.

Chapter Ten REVIEW

1. Define solid modeling.
2. Define wireframe modeling.
3. Define surface modeling.
4. List the two common methods of creating solid models.
5. List the four design-related functions that can be performed on a 3-D model.
6. List four different methods of creating wireframe models.
7. List some common mass property calculations that can be determined with CAD.

Chapter Ten DRAWING EXERCISES

1. Draw the V-Block from Chapter Six in isometric and oblique.
2. Create Wireframe models of Figures 10-39 through 10-45.

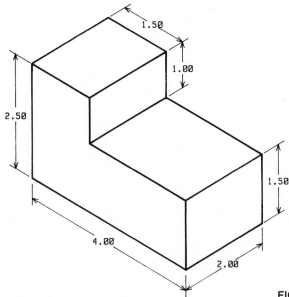

FIGURE 10-39   Corner block

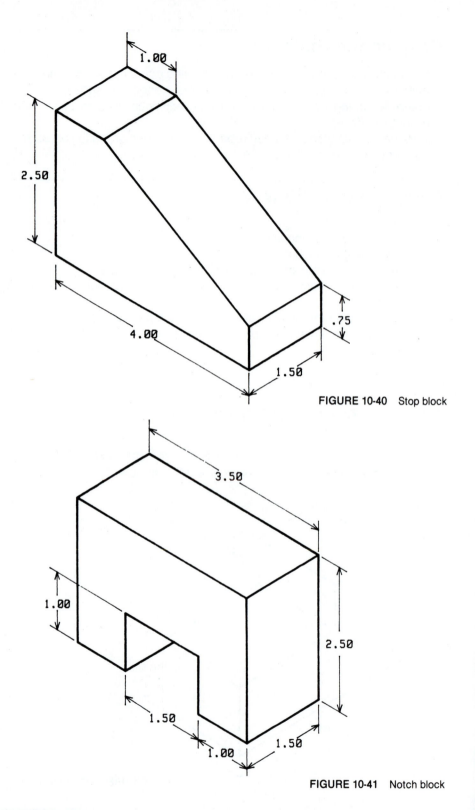

FIGURE 10-40 Stop block

FIGURE 10-41 Notch block

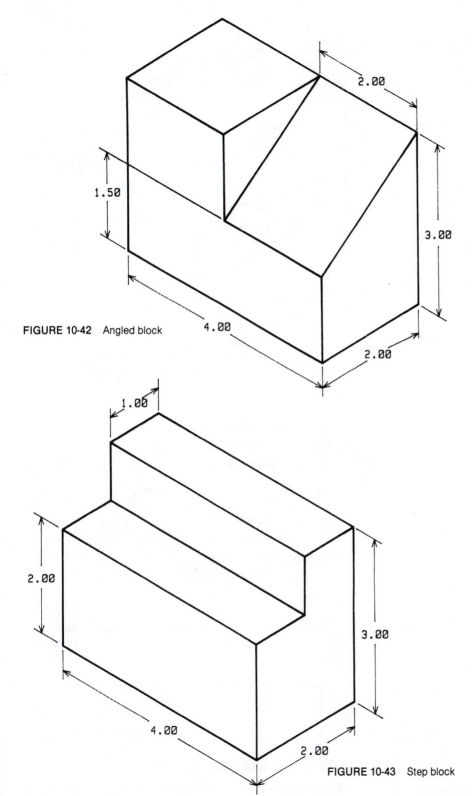

FIGURE 10-42   Angled block

2.00

1.50

3.00

4.00

2.00

1.00

2.00

3.00

4.00

2.00

FIGURE 10-43   Step block

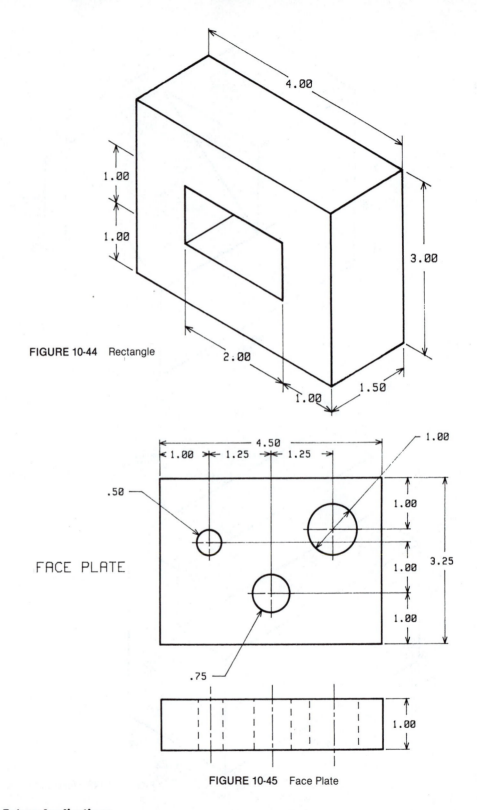

FIGURE 10-44 Rectangle

FACE PLATE

FIGURE 10-45 Face Plate

3. Follow the instructions outlined in caption for Figure 10-46.

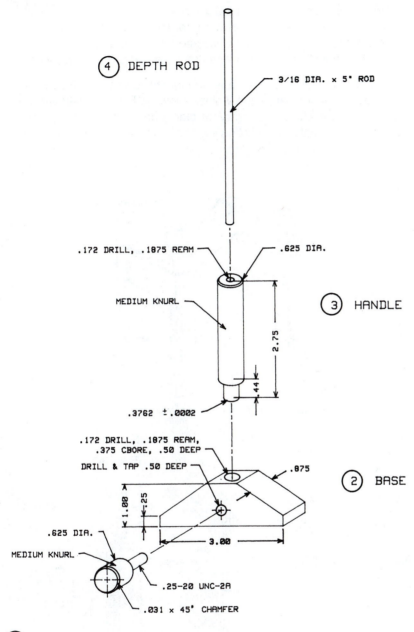

4 DEPTH ROD

3/16 DIA. x 5" ROD

.172 DRILL, .1875 REAM

.625 DIA.

MEDIUM KNURL

3 HANDLE

2.75

.44

.3762 ±.0002

.172 DRILL, .1875 REAM,
.375 CBORE, .50 DEEP

DRILL & TAP .50 DEEP

.875

2 BASE

1.00

.25

.625 DIA.

MEDIUM KNURL

3.00

.25-20 UNC-2A

.031 x 45° CHAMFER

1 1" LONG KNURLED SCREW

FIGURE 10-46 1. Make a working drawing of each part of this Depth Gauge.
2. Make an assembly drawing with a parts list.
3. Make an enlarged section through assembly of the base.
4. Use different pens to improve the clarity of the drawings.
5. Use layers to separate entities.
6. Make plots of all drawings.

4 . Following are instructions for Figures 10-47 to 10-49:

1. Create a working drawing of the V-block.
2. Create a pictorial assembly drawing for technical illustration.
3. Draw threads in working drawings and pictorial drawings.
4. Draw sectional views on working drawings and pictorial drawings.

To accomplish this project, you will refer to Figures 10-47 to 10-49 and make a working drawing of each part of the V-block shown in Figures 10-48 and 10-49. Add sectional views where necessary for clarity and dimensioning purposes. Draw a detailed pictorial assembly drawing similar to that shown in Figure 10-47 for a technical illustration.

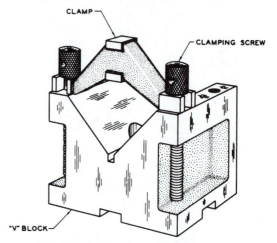

FIGURE 10-47   V-Block assembly drawing.

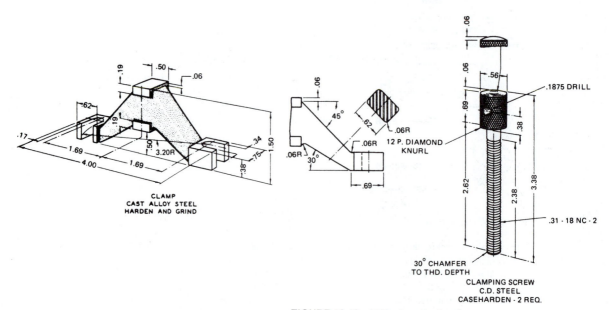

FIGURE 10-48   V-Block parts drawing.

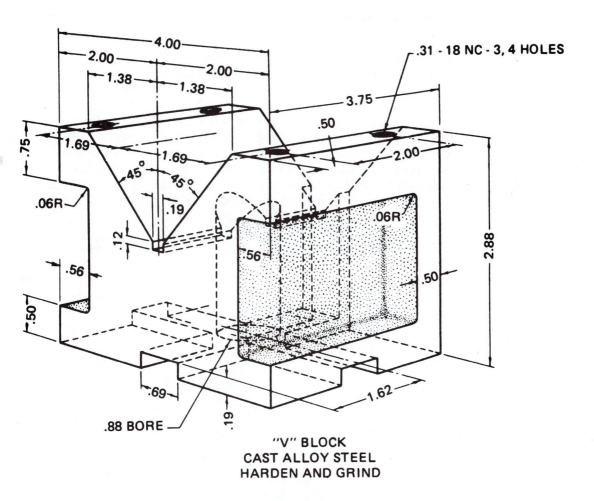

"V" BLOCK
CAST ALLOY STEEL
HARDEN AND GRIND

FIGURE 10-49   V-Block details.

5. Following are instructions for Figure 10-50:

   1. Create a solid model of the Bell Roller Support.
   2. Create a cutaway section of the model.
   3. Create orthographic drawings of the model.
   4. Create an NC program for the Base Plate and Shaft.

To accomplish this project, you will refer to Figure 10-50 and design the Bell Roller Support Assembly as shown. This assembly is to support a 1-inch flat belt to run at 45 revolutions per minute. The approximate dimensions of the base plate are 2 inches by 3.5 inches and the bell roller diameter is approximately 3 inches. Make a model of each part. Create an isometric exploded assembly drawing for a technical illustration. Create an NC program to turn the shaft on a lathe and to mill the base plate.

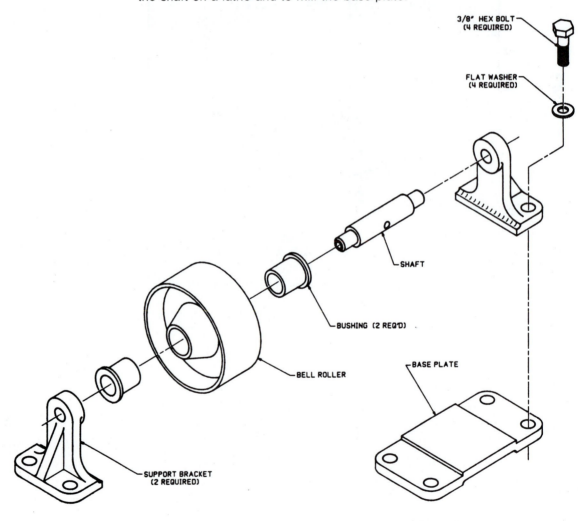

FIGURE 10-50   Bell Roller Support assembly. *(Courtesy of International Business Machines Corporation)*

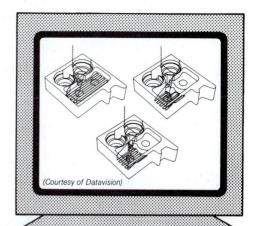

(Courtesy of Datavision)

This chapter includes a discussion of the magnitude of change that is expected in manufacturing and society because of the computer. You will learn how CAD fits into the overall manufacturing process, and how the computer will be used to link all phases of the manufacturing operation. You will also learn about the role that robots will have in the automated factory.

The use of computers, and the drive to produce an automated factory run by a small number of technicians, will have a profound effect upon manufacturing, the worker, and society. Most people do not realize the magnitude of change that will come about because of factory automation. The changes are expected to be so great that the advent of computer-controlled automated factories is being called the second industrial revolution. These changes are so far-reaching that few people know yet where this movement will lead and what its overall effects will be.

This chapter gives you some insight into this second industrial revolution. You will gain an understanding of how CAD and CAM can be brought together to enhance the manufacturing process. The concept of the Flexible Manufacturing System (FMS) using robots is explained. Computer Integrated Manufacturing (CIM) is also covered, showing how computers can be used to control the entire business operation, from placement of the order to shipment of the final product and all the steps in between. Finally, the effects of the automated factory on workers and society is explained.

Chapter Eleven
CAD/CAM, Applications, and the Factory of the Future

OBJECTIVES

After completing this Chapter, you will be able to:
- Define CAD/CAM and explain its function.
- Define CAM.
- Define NC, CNC, and DNC.
- Define FMS and explain how it operates.
- Define CIM and explain how it operates.
- Explain the role of robots in automated manufacturing.
- Describe some of the effects of factory automation on society.
- Describe some applications of CAD.

A SCENARIO OF THE FACTORY OF THE FUTURE

The factory of the future may be radically different from the factory we know today. The following scenario or future projection will help you to visualize how the factory of the future will operate, the control that computers will have in manufacturing, and the limited extent of human intervention. In story form, the scenario follows.

Bryan, the recently hired designer, quickly finished his breakfast so that he could complete his design of the turbine blade of a new steam turbine generator. He went to the office in his home where the CAD workstation was located. Putting on the headset to communicate orally with the workstation, Bryan called up the 3-D view of the turbine blade. The blade quickly appeared on screen. Bryan began to rotate it to view the blade at various angles. He zoomed in on the end of the blade to complete the detail work needed to attach the blade to the hub of the turbine assembly.

After completing the detail work on the end of the blade, Bryan called up the hub assembly from the steam turbine generator data base from the main computer bank located more than one thousand miles from his home. The hub was created by a designer named Ada. This was all that Bryan knew of his colleague who designed the hub. The hub quickly appeared on screen, and Bryan moved the turbine onto the hub to check the fit. Satisfied that the design would fit within the specified tolerance, he quickly made a symbol of the turbine blade and located nine more blades around the hub to complete the design.

Bryan then ran stress-strain analyses on the design. The assembly quickly filled with colors of blue green, yellow, orange, and red. The colors represented the amount of stress on the assembly—with blue being the lowest stress, to bright red representing maximum stress. Bryan chose a special alloy material for the blade assembly that would withstand the stresses shown on screen.

He then called up the entire generator assembly to check his design. The actions of the turbine assembly were then simulated on screen for one final check of his design. Finally, he was ready to beam his design to the satellite that would relay the information to the assembly plant overseas in Norway.

The design was received within seconds at the terminal of Kevin, the manufacturing engineer. Kevin called the assembly up on his screen as a final check. The design was then sent to the main plant computer that controlled all manufacturing operations. The computer made up the number of turbine assemblies needed and sent this order to supplies. As the order was sent, an available flexible manufacturing system (FMS) was found and the robots and machine tools were programmed from the design data base. The material orders were sent to supply, and the special alloy material specified in the design was delivered on automatic carts to the FMS where robots handed them to the machine tools. The finished turbine blades were then transferred to the assembly area where robots put the blades on the turbine generators. Sensor robots monitored the entire manufacturing process and performed quality control inspections. The finished assembly was delivered to shipping on automatic carts and placed by robots into special shipping containers. The factory runs 24 hours a day, 7 days a week, with just a skeleton crew of engineers and technicians.

THE AUTOMATED FACTORY

This scenario may seem like science fiction to those not familiar with the research and the advances expected to take place with computers and robots. Actually, many of the operations described can be easily accomplished with today's technology. What this scenario suggests is a computerized, paperless, workless manufacturing system that will shake the very foundation of our industrial society. Approximately 20 percent of our work force is engaged in manufacturing. This number is sure to decline as automation of factories increases. Our slide from a manufacturing society to an information or service society will continue. This movement is being called the "second industrial revolution."

CAD/CAM

Computer-aided Design/Computer-aided Manufacturing (CAD/CAM) is an umbrella process that combines the design and manufacture of a product into one integrated approach. Computer-aided drafting is only one part of the overall CAD/CAM process. Using CAD to produce mechanical drawings is the major function of many CAD systems. However, the trend in computer graphics is beginning to include CAD systems that can do more than just make orthographic drawings. Many systems on the market are referred to as CAD/CAM systems. With this type of system, drafting is only a part of the overall capabilities.

In general, CAD systems provide the data base needed to manufacture a product. The importance of this data should not be underestimated. Traditionally, designing and manufacturing have been separate activities. CAD/CAM can provide a direct link between these two activities. The manufacturing data base can become part of the CAD/CAM data base. Much of this data is produced in the initial design of the part using CAD. The advantages of having access to this data include:

1. Numerical control machine tool programming
2. Tool and fixture design
3. Computer aided process planning (CAPP)
4. Direct numerical control of machine tools (DNC)
5. Computer-aided inspection
6. Robotic and automated vehicle planning
7. Group technology (GT)
8. Automation of the product from receipt of the order to delivery
9. Development of flexible manufacturing cells (FMS)

The Role of CAD in the Automated Factory

The role of CAD in the factory of the future is already firmly established in the design end of the product and is expected to grow approximately 30 percent a year. With CAD, the product can be designed using orthographic views or in 3-D using wireframe or solids modeling. See Figure 11-1.

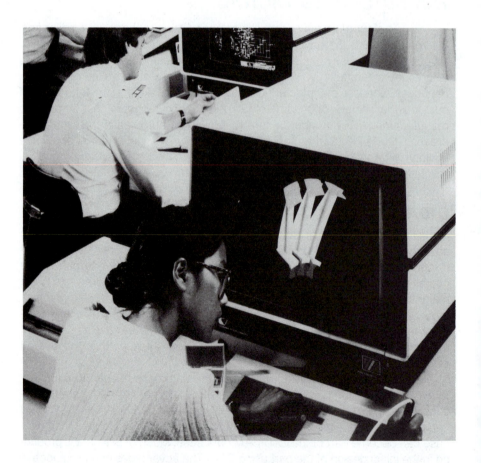

FIGURE 11-1 A CAD/CAM system displaying a solid model. *(Courtesy of Computervision Corp.)*

Modeling allows colored surfacing with varied intensity values for shading and creating different surface textures. The choice of colors is virtually limitless, with some systems offering the designer a palette of over one million colors to choose from when creating a model. It should be noted that, because of limited memory, the number of colors that can be displayed at one time on screen is considerably less than one million. The designer can "walk" inside and redesign interiors of models. Mass properties, such as weight, volume, area, center of gravity, and moments of inertia can be determined. The designer also has the ability to make an assembly fly apart. The designer can analyze the static and dynamic properties of a design. Design changes and the investigation of alternative designs can be done easily. Models can be dynamically displayed on screen showing how parts interact without the need for building prototypes.

The Role of CAM in the Automated Factory

Of course, the role of computers does not stop with the design of the product. The manufacturing of the part can be controlled by computers. This process is called computer-aided manufacturing (CAM). CAM had its beginning with the numerical control (NC) of machine tools. *Numerical control (NC)* tools are automated machines controlled by programs, punched on paper tape or magnetic tape used to fabricate material. For more than 20 years, NC machine tools have been controlled by the use of punched tape to drill, cut, mill, punch, weld, and grind raw materials. These punched tapes control the machine tool operation so that identical parts can be produced rapidly and accurately. Recently these NC machines have been controlled by computers in a process called computer numerical control. *Computer numerical control (CNC)* is the use of a computer to control some or all of the NC functions. Figure 11-2 shows a CNC machine tool. When a group of NC or CNC machine tools is controlled by a host computer, the process is called *direct numerical control (DNC)*.

Computer-aided manufacturing (CAM) can mean anything from automated machine tools to robots programmed for specific tasks. CAM is used when a number of identical parts needs to be fabricated or assembled. Because each part is identical, a program can be written to control machine tool movements producing parts exactly alike. Writing and debugging NC programs is tedious and expensive. For these reasons, an alternative to writing programs has evolved. Traditional methods of programming NC machine tools were manual part programming, or writing a sequence of commands to control the movement of the tool, and computer-assisted part programming. Interactive graphics can now be used to assist in programming machine tools. Software has been produced that uses the geometric data base produced when the part was designed with CAD to make the NC program. This joining of CAD and CAM is the key to producing an automated factory.

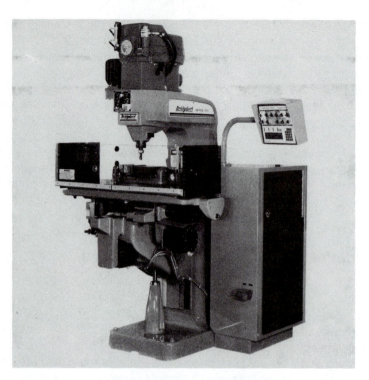

FIGURE 11-2 A CNC milling machine. *(Courtesy of Bridgeport Machines)*

Computer-aided Manufacturing

With the product designed and documented you are ready to manufacture the part. *Computer-aided manufacturing* (CAM) automates the manufacturing operations using computer-controlled machines and robots for assembling and handling materials, measuring, and inspection. The first step in CAM is to generate the numerical control (NC) program to machine the parts. This is done by using the geometry created when designing the part. The parts programmer uses this data base to interactively create the tool path with up to five axes to machine the part. A library of machines and tools is usually available to the programmer for developing and editing the tool path. After the program has been developed, the programmer can check the tool path on screen by simulating the machining operation. See Figure 11-3. This dynamic display will show the cutter and/or the cutter path center line. After the programmer makes any necessary changes in the program, it is either downloaded to the NC machine or a tape is punched for machining later. See Figure 11-4.

Most CAD/CAM systems are capable of performing the functions just described. As can be seen, the computer is even automating the creation of orthographic drawings by automatically making the desired views from the 3-D model. Parts programming for NC machining is also becoming more automated with CAD/CAM. The result is increased productivity, better designs, and decreased time from design to manufacturing.

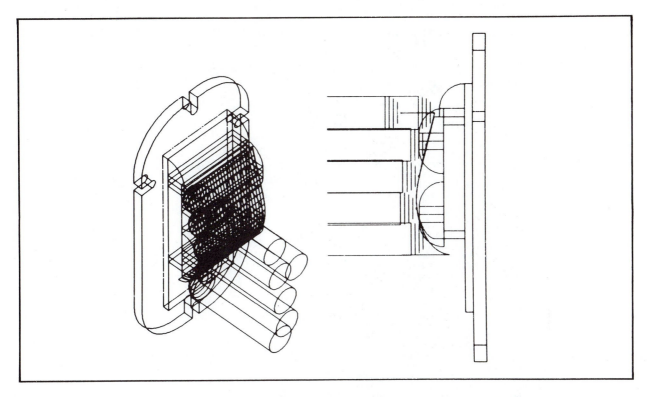

FIGURE 11-3 Simulation of the NC tool path of a ball end mill. *(Courtesy of Calma Company)*

The CAD/CAM Process Applied

CAD/CAM is the process of combining NC machines, CAD, and manufacturing resource planning into one integrated approach. This integrated approach is explained using Control Data's ICEM CAD/CAM system as an example. Refer to Figures 11-5 through 11-12.

A typical CAD/CAM system creates a geometric model of the part to be designed with a menu-driven, interactive modeling package. The model is rotated, exploded, and viewed from any angle. A finite element model is generated and analyzed to check design specifications. The designer or NC programmer then determines the tool path needed to produce the part. The tool path can then be dynamically displayed on screen over a 3-D model of the part to visually check the program. The program is then translated into the machine language of the CNC tool by a postprocessor. A *postprocessor* is a computer program used by the processor to change or convert data to another form.

In this example, the geometric data of the part design and the tool path shown on the CAD terminal is converted into a program that can be used by an NC tool. Different postprocessors are used to convert this data to various machine tools, such as milling machines, lathes, drills, and so

```
$ SYSTEM DATE      13-Jun-85
$ SYSTEM TIME       5:31
$ PARTNO GASKET
MACHIN,
SETUP,
$NO DIMS
DLN1,PT(-0.1507XA,1.7407YA),PT(-0.3667XA,1.4593YA)
DCIR1,0.0000XA,1.6250YA,0.1900R
DLN2,PT(1.3409XA,-0.5926YA),PT(0.8195XA,-1.7722YA)
DLN3,PT(0.7043XA,-1.7213YA),PT(1.2256XA,-0.5417YA)
DCIR2,-0.5253XA,1.5810YA,0.2000R
DCIR3,0.5253XA,1.5810YA,0.2000R
DCIR4,-0.6408XA,0.8817YA,0.2500R
DCIR5,0.0000XA,0.0000YA,1.3400R
DLN4,PT(0.3667XA,1.4593YA),PT(0.1507XA,1.7407YA)
DLN5,PT(-0.8905XA,0.8687YA),PT(-0.7689XA,-1.4503YA)
DCIR6,1.1065XA,1.2455YA,0.2000R
DCIR7,1.6319XA,0.3355YA,0.2000R
DCIR8,1.4073XA,0.8125YA,0.1900R
DLN6,PT(1.4321XA,1.0009YA),PT(1.0804XA,1.0472YA)
DLN7,PT(1.4471XA,0.4121YA),PT(1.5828XA,0.7398YA)
DCIR9,0.0000XA,0.0000YA,1.4660R
DCIR10,0.0000XA,0.0000YA,1.4660R
DCIR11,0.0000XA,0.0000YA,1.4660R
DCIR12,-1.4170XA,1.3628YA,0.5000R
DCIR13,-1.3647XA,0.3642YA,0.2500R
DCIR14,0.0000XA,-1.4100YA,0.7700R
DCIR15,-2.2524XA,-0.8939YA,0.3079R
DLN8,PT(-2.0444XA,1.0144YA),PT(-2.6810XA,0.3777YA)
DCIR16,-1.9100XA,0.8800YA,0.1900R
DCIR17,-2.5042XA,0.2009YA,0.2500R
DCIR18,-2.2354XA,0.1162YA,0.2500R
DCIR19,-1.4170XA,1.3628YA,0.7500R
DCIR20,-1.8768XA,0.4748YA,0.2500R
DCIR21,-0.3772XA,-2.4390YA,0.2000R
DCIR22,-2.1542XA,-0.3447YA,0.2500R
DLN9,PT(-2.7504XA,0.1575YA),PT(-2.3871XA,-1.9030YA)
DLN10,PT(-2.0536XA,0.6515YA),PT(-2.4122XA,0.2929YA)
DCIR23,0.0000XA,-1.4100YA,0.8960R
DCIR24,0.0000XA,-2.3400YA,0.1900R
DCIR25,0.3772XA,-2.4390YA,0.2000R
DCIR26,-0.7094XA,-2.3100YA,0.2500R
DLN11,PT(-2.2000XA,-2.0600YA),PT(-0.7094XA,-2.0600YA)
DCIR27,-1.7885XA,-1.4754YA,0.2500R
DCIR28,-1.2635XA,-1.5680YA,0.2500R
DLN12,PT(-2.4816XA,0.0728YA),PT(-2.4004XA,-0.3882YA)
DLN13,PT(-1.9493XA,-0.9473YA),PT(-2.0347XA,-1.4320YA)
DLN14,PT(-1.8319XA,-1.7216YA),PT(-1.3069XA,-1.8142YA)
DLN15,PT(-1.0138XA,-1.5549YA),PT(-1.1150XA,0.3772YA)
DCIR29,-2.2000XA,-1.8700YA,0.1900R
DCIR30,0.0000XA,-1.4100YA,0.8960R
DPT1,0.0000XA,1.6250YA
DPT2,0.0000XA,-2.3400YA
DPT3,1.4073XA,0.8125YA
DPT4,-1.9100XA,0.8800YA
DPT5,-2.2000XA,-1.8700YA
DPT6,-0.8721XA,-1.8700YA
DPT7,-2.2524XA,-0.8939YA
DCIR31,-2.2524XA,-0.8939YA,0.3079R
```

FIGURE 11-4 An NC drawing of a gasket with the NC program listing generated from the data base.

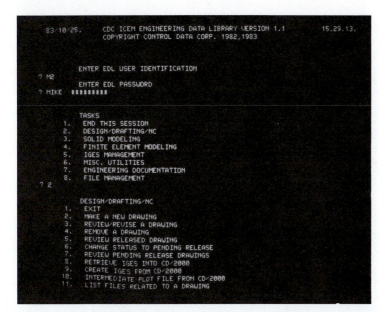

FIGURE 11-5 Task selection menu from Control Data's ICEM (Integrated Computer-Aided Engineering and Manufacturing) Engineering Data Library (EDL). EDL manages engineering data, and makes it available to all functions. *(Courtesy of Control Data Corp.)*

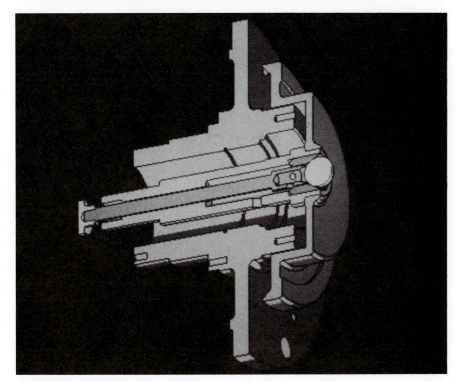

FIGURE 11-6 Control Data's ICEM solids modeler helps create, analyze, and visualize the geometry. *(Courtesy of Control Data Corp.)*

FIGURE 11-7   Solids model of the assembly can be rotated, exploded, and viewed from any angle. *(Courtesy of Control Data Corp.)*

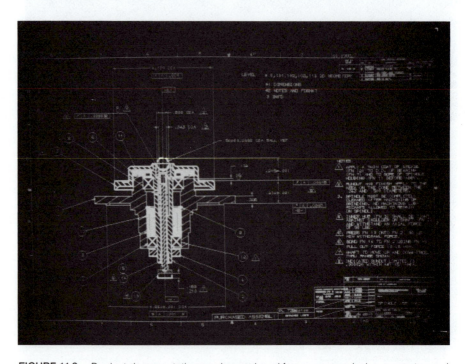

FIGURE 11-8   Product documentation can be produced from common design geometry, and maintained and retrieved by the system. *(Courtesy of Control Data Corp.)*

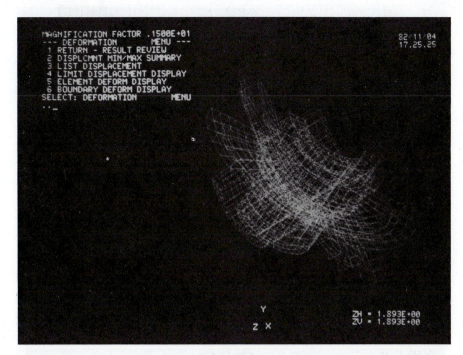

FIGURE 11-9   A finite element model is generated from common geometry and analyzed to ensure meeting design criteria. *(Courtesy of Control Data Corp.)*

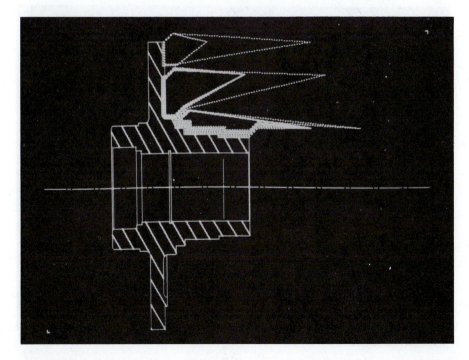

FIGURE 11-10 Cutter paths for numerical control machining can be defined and modified. *(Courtesy of Control Data Corp.)*

FIGURE 11-11 Numerical control output can be used for NC machining of the actual part. *(Courtesy of Cincinnati Milacron)*

FIGURE 11-12 Finished assembly view. *(Courtesy of Control Data Corp.)*

forth. The machine program can then be punched on tape, stored on disk or downloaded directly to the tool for immediate part fabrication. The capability of programming robots on CAD/CAM systems is also being developed. The actions of a robot can be simulated on a CAD system to significantly ease the writing of the program.

The advantages of CAD/CAM include increased productivity, design of more sophisticated projects, less design time and more design alternatives. Parts can be checked on screen for fit, and the designer can check to see if a part already exists in the main computer. CAM tools and robots can complete the entire manufacturing sequence from design to manufacture of the part. The data bases created by a CAD/CAM system can be sent over phone lines to other manufacturing plants.

The Flexible Manufacturing System (FMS) or Cell

The *flexible manufacturing system (FMS)* or cell is a series of computer-controlled machines that perform a complete task and are serviced by computer-controlled transfer devices such as robots. Figure 11-13 shows an FMS cell used to automatically fabricate parts. The cell is automatically supplied with raw materials and automatically transfers finished parts with very little human intervention. Many such cells are already in use throughout the world. Japan is working on a large government project having as its goal an unmanned factory. Already existing in Japan is a factory that assembles vacuum cleaners without human aid. The factory is left virtually unattended except for a handful of technicians who are there for part of the day.

Robots

A major part of an automated factory is the use of robots. A *robot* is a programmable device used to move and fabricate parts and materials. Figure 11-14 shows a robotic arm used for plasma cutting of metal. Robots are used where repetitive or dangerous operations must be performed. Because the sequences of steps are identical, a program can be written to control a robot's motions. The use of robots in industry is already firmly established, and is expected to take over more repetitious and dangerous jobs performed by humans.

Robots can be programmed in three different ways. The *Walk-through* method involves the programmer manually moving the robot's arm and hand through the work cycle. This movement is recorded in the robot's controller memory. The *Lead-through* method uses a teach pendant that the programmer uses to remotely control the robot's movement through the work cell. The movement is recorded in the robot's memory for later use. *Off-line programming* can be compared to NC programming. The program is written on a computer terminal and down-loaded to the robot. With this technique of programming robots it is possible to use the CAD/CAM data base to assist in the generation of the program. The robot can be graphically simulated on screen along with the com-

FIGURE 11-13 Overall view of the Kearney & Trecker flexible manufacturing system (FMS) installed at Hughes Aircraft Company, Electro-Optical & Data Systems Group, El Segundo, California. *(Courtesy of Kearney & Trecker Corp.)*

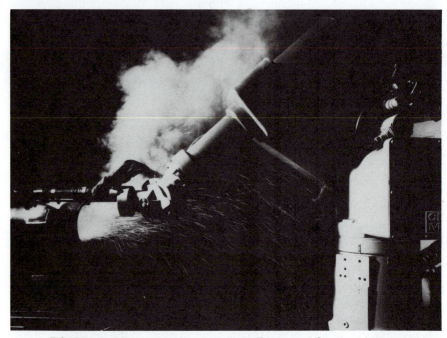

FIGURE 11-14 Plasma cutting robotic arm. *(Courtesy of Cincinnati Milacron)*

ponents in its work cell. Typical components could include NC machine tools, conveyor belts, fixtures, coordinate measuring machines, and so forth. The CAD system could be used to position the work cell components and simulate the robot's movements.

Computer Integrated Manufacturing (CIM)

CAD/CAM and FMS deal only with the automation of the design and manufacture of a part. *Computer integrated manufacturing (CIM)* is the total automation and computerization of the manufacturing process from start to finish. CIM automates all facets of the manufacturing process from the orders, through design, scheduling, materials handling, manufacturing, inspection, assembling, and maintenance to the shipment of the final product. A typical part takes only about 5 percent of the time being machined and 95 percent waiting or moving. The productivity improvement in this area alone can be significant with CIM. CIM can increase productivity, improve cost, produce better products, lower inventories, and improve accounting and purchasing procedures.

With CIM, a common pool of data is available to all those working in the factory. The creation of a common data base for all facets of a factory is the key to CIM. Figure 11-15 shows CASA's (Computer and Automated Systems Association of SME) graphic representation of CIM. The wheel shows how all facets of a manufacturing operation are integrated and have access to common data through the use of computers. Using CAD/CAM or CIM to its greatest capacity means using the computer for all facets of a manufacturing operation.

Looking at the large inner wheel in Figure 11-15, many of the departments and activities involved with CIM are shown. Any information or design output by one department becomes the input for another department. For example, the output of a part design becomes the input necessary for analysis and simulation. One advantage to the designer with CIM is that previously designed parts can be found from the common data base, thus eliminating repetitious designs. This is commonly referred to as *Group Technology* (GT). *Part Families* can be developed to group parts which are similar in geometric shape or manufacturing methods. The departments or modules are controlled by computers that are all interconnected.

CIM improves productivity, lowers labor costs and product costs, improves quality, and changes the skills needed for employees. Repetitious and dangerous jobs will be eliminated because of automation. Eliminating such jobs will require trained and skilled programmers, technicians, and maintenance workers. The consequences of widespread automation on the economy and society are difficult to predict. Will this electronic revolution cause the far-reaching problems as did the first industrial revolution?

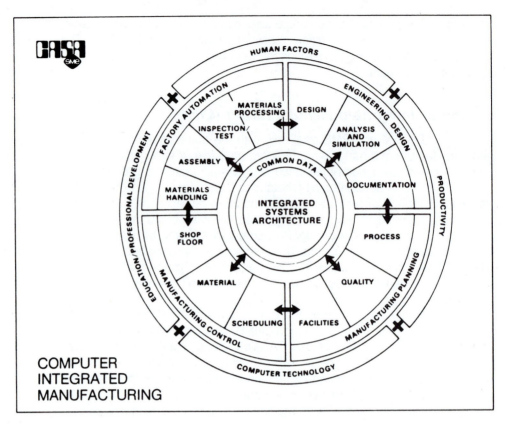

FIGURE 11-15 Graphic representation of computer integrated manufacturing. *(Courtesy of CASA/SME)*

THE FIRST INDUSTRIAL REVOLUTION

Writing in 1884, the historian Arnold Toynbee popularized the term *industrial revolution* to characterize the massive transformation of England in the late eighteenth and nineteenth centuries from an agricultural and commercial society to an industrial society, including major social and political changes. This revolution resulted in violent changes which brought about an age of economic exploitation and social unrest.

The revolution started with the use of steam engines to drive the machinery of textile mills in the late eighteenth century in England. This led to the crowding of thousands of families into and around cities having textile mills. The cottage approach to industry was largely abandoned as factories proved to be more economical and productive. Workers were expected to work long hours and child labor was common. Poor and unsafe working conditions were other problems created by the revolution. These problems saw little improvement until the mid 1800s.

To summarize, the first industrial revolution produced some of the most complete changes to occur in history, resulting in significant political and social changes. The second industrial revolution promises more of the same if nations do not prepare for the consequences of a computerized society. Using the full power of the computer will have one of its greatest effects on the basic manufacturing industries of our country which account for approximately 20 percent of the total work force.

THE SOCIAL AND ECONOMIC CONSIDERATIONS OF THE AUTOMATED FACTORY

The first revolution brought about the crowding of people into cities close to the factories. The second industrial revolution will see an increase in the flight from the cities that is occurring today. This exodus will likely lead to the decay of some large cities. Communications by phone or satellite will allow companies to disperse operations to smaller cities closer to large markets.

Jobs should become less toilsome and dangerous because of automation and the use of robots. Because of computers, there will also be an increase in the number of cottage industries and the number of jobs that can be performed at home. Our society will change from smokestack industries to electronics industries, and from a manufacturing base to an information age. It is estimated that approximately 86 percent of the labor force will be involved in making computers, providing computer services, and distributing computer information to society.

Government will take an increased role to provide more services and protect people from possible abuses of the computer. There will be a large displacement of workers as the economy changes from a manufacturing society to one of service and technology. It is estimated that by the year 2000 manufacturing jobs will account for only about 10 percent of the work force compared to 20 percent now. Service-related employment will increase from 60 percent to 86 percent.

As technology and knowledge increase at a dramatic rate, displaced workers and graduating high school students will need to be trained for available jobs. These jobs will include: computer programmers, robotic technicians, electronics technicians, telemarketing, and CAD/CAM. Education and job training will become a lifelong process as the technology of the job changes.

The increased use of computers in society and manufacturing raises many questions. In the past, automation produced more jobs than were displaced. Will this trend continue? What should be done about the displaced workers? Because factories will become more efficient, and thus more profitable, will this increased profit be shared with the workers? What other societal problems will arise? These and other questions need to be answered.

The first industrial revolution produced major unplanned changes in the environment and society, and for the worker. The second industrial revolution, led by the computer, should be studied and planned for in advance. Studies should be made on the long range effects of computers and automation on factory workers and society. The second industrial revolution is upon us. Common sense and empathy must be the indispensable difference between the first industrial revolution and this electronic revolution and resultant upheaval of the worker.

Are we prepared for this revolution or will the revolution catch us unprepared? CAD training is only the first step in a lifelong commitment that a person must make to stay current with the dynamic technology associated with a career in drafting/design.

APPLYING CAD TO MECHANICAL DRAWINGS

CAD provides the user with a flexible, easy to use, automated method of producing mechanical drawings. Mechanical drawings are used for fabrication of materials, tool and die design, jig and fixture design, design of molds, machine design, and product design. The many benefits of using CAD have been covered in preceding chapters. You will find that the skills learned, combined with the full capabilities of the CAD system, will make the creation of mechanical drawings easier and faster than using traditional tools.

Before starting on the drawing, there are a number of points to be considered by the operator. The following is a list of some of these points to consider before beginning on the drawing to gain maximum use and efficiency from the CAD system.

- What name and part number are to be assigned to the drawing?
- Choose the paper size and the scale to be used on the drawing.
- Choose the dimensioning style and units.
- Decide which layers will be assigned to entities.
- Will you need to create an automatic bill of materials?
- Assign pen numbers to various entities.
- Can any current symbols be used, or will you create your own symbols for this drawing?
- Can a parametric program be used for any part of the drawing?
- Are any cutaway or special views needed?
- Sketch or have an idea of the views and the layout to be used.
- Will any shortcuts using CAD be used, such as mirroring or copying?
- Think carefully through the steps you are going to follow to create this drawing using CAD.

After these considerations have been resolved, you can start on the drawing confident that your preplanning will eliminate most errors and the necessity for redrawing. The color insert explains some applications of mechanical engineering using CAD.

APPLYING CAD TO ARCHITECTURAL DRAWINGS

Most CAD software packages can be used for architectural drawings. However, some software packages are made specifically for architectural drawings. These software packages have many shortcuts and special features to assist the user in developing architectural working drawings. Figure 11-16 shows an architectural drawing produced with a standard drafting package.

Before starting on your drawing, you must gather information necessary for the design of the structure. Style and construction, location on the lot, financial considerations, and any special needs of the occupants must be considered before drawing. After you have this information, preliminary sketches of the structure are made. After you are satisfied that the sketches meet the client's needs, you can start planning the use of the CAD system for drawing the working drawings.

Before using the CAD system to draw, the following important considerations should be reviewed.

- Determine the paper size for final plots.
- Determine the scale to be used on each drawing.
- Assign layers to the various drawings to be created. For example, the floor plan can be used to lay out the basement and foundation plans.
- Assign pens to various drawing entities. The use of color can greatly enhance the legibility of the drawing and improve the overall appearance.
- Choose the dimensioning style and units.
- Determine the symbols to be used or create any needed symbols.
- Sketch a layout of the drawings.
- Determine if there are any shortcuts that can be made using CAD. For example, the floor plan without dimensions can be used for the electrical plan, saving the operator considerable time, and the COPY command can be used to copy details on elevations.
- Choose a lettering style.
- Use the "bill of materials" function to create window and door schedules.
- Determine carefully the steps to follow to create each drawing of the set of working drawings.

Through careful preparation, the operator can save a significant amount of time and make full use of the capabilities of the CAD system. It is important to remember that a CAD system is only as good as the person operating it. If operators fail to use the various shortcuts and full capabilities of the system, they may be better off using traditional tools to create drawings. CAD is a costly investment, and it only becomes cost effective when it is used to its full capabilities. Planning ahead and determining how the drawings are to be created will ensure that CAD is being used to its full capabilities.

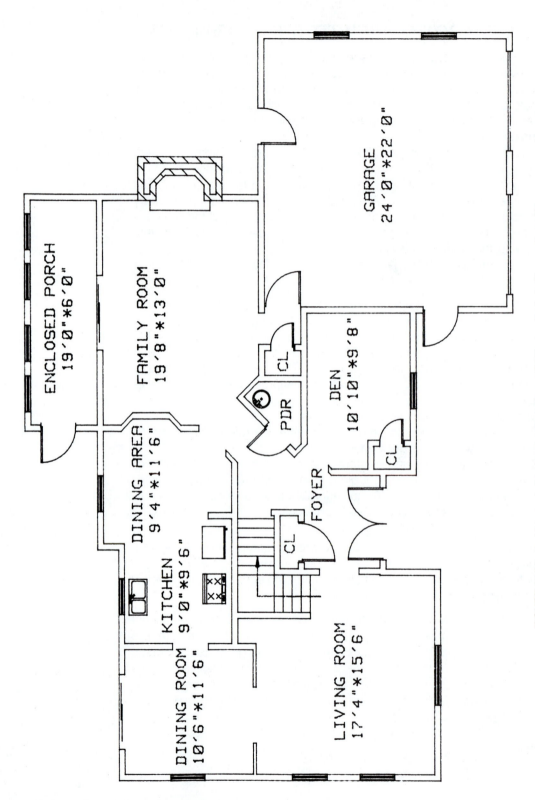

FIGURE 11-16 Architectural drawing made using a CAD standard drafting system. Doors, windows, and appliances were made using symbols.

Architectural Software Packages

Any standard CAD drafting package can be used to create architectural drawings. These standard drafting packages enable the operator to create architectural drawings faster and more neatly than by using traditional tools. Special software packages have also been developed specifically for the architectural drafting market. These packages have special drawing features that allow architectural drawings to be created faster and easier than with standard drafting packages. Some of these special drawing features, described next, are only a sample of some of the capabilities offered in architectural software packages.

Capabilities of Architectural Software

Walls can be drawn by identifying the location of the center line and its length. The wall thickness is automatically drawn to a length specified by the user. Doors can be located in walls by identifying the center point and the direction of the swing. The wall is automatically clipped, and the door is drawn in the wall showing the direction of swing. While designing rooms, the operator can dynamically move or change the room size on screen. After the drawings have been created, the operator can extract schedules or square footage of rooms. Interference checks between plumbing and heating, for example, can be done by overlaying these two plans and checking for any problems. Details can be used on other drawings by editing changes. Pattern fills, such as brick, block, earth, and so forth, can be easily placed on drawings.

Additional Architectural Application Programs

Architectural modeling is a very important part of architectural drawing. With a modeling package, the user can create a 3-D model and "walk" through the model by using the ZOOM command and changing the viewing point. The outside of the building can also be viewed from numerous vantage points. Space planning is another powerful tool that can be used to complete layout drawings efficiently showing room components and producing reports. See the color insert for other examples.

ELECTRICAL DRAWINGS

Most standard CAD drafting packages are capable of producing electronic schematic drawings, block diagrams, and wiring diagrams. Using CAD to produce any one of these drawings greatly enhances the user's ability to produce, save, recall, and place symbols. Schematic symbols, blocks, rectangles, triangles, and circles can be easily produced with a CAD system and saved as a symbol for repeated use. For these and other reasons, standard CAD drafting packages can be used efficiently for electronic drawings. With CAD, electronic drawings are produced much more easily than they are by using traditional drafting practices.

Drawing Schematic, Block, and Wiring Diagrams with CAD

As with any drawing, a number of procedures must be followed before the final drawing begins. Most designs begin with a sketch of the circuit. Before drawing the circuit using CAD, the following factors should be determined.

- Choose the paper size and scale.
- Choose the lettering style.
- Decide which layers will be assigned to different entities.
- Will you need to create a bill of materials?
- Assign pens to different entities.
- Determine the symbols to be used and which ones, if any, must be drawn.
- Think carefully through the steps you are going to follow to create the drawing.

Once these considerations have been determined, you can begin drawing. The use of color in creating electronic drawings is very helpful in distinguishing different circuits and components. So, consider the use of color before you begin to draw. Figure 11-17 shows an electronic schematic drawing produced on a CAD system. The original of this drawing shows the callouts in red.

ELECTRONIC DESIGN PROGRAMS

Some of the same types of automated activities described previously for producing mechanical drawings can also be applied to electronic design. Automatic routing and checking of circuits on PC boards is possible with a special application program on some CAD systems. Drill tapes for NC machines can also be generated automatically after the design is completed. Output of photoplots for screening the PC board, and pen plots for checking circuit design can be created from the post processor. Automatic assembly of the PC board can also be accomplished using robots and automatic machinery. See color insert for example PC Board and Electronic Circuit Design.

BUSINESS GRAPHICS, GRAPHS AND CHARTS

Another major application area for CAD is in the producing of graphs and charts. Graphical display of information is much easier to understand than the simple listing of numbers. For this reason, graphs and charts are used extensively in business and industry. CAD makes the development and display of graphs and charts easy and fast. Because plots and display devices can be in multicolor, CAD can produce graphs and charts that are much easier to read and understand. The CAD operator should keep in mind that colors used with different fill patterns can greatly enhance a graph

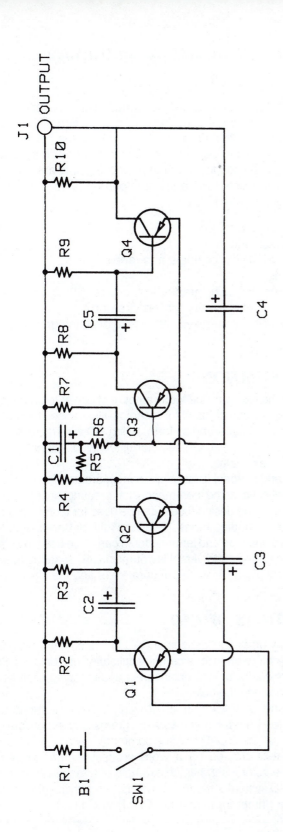

FIGURE 11-17 This simple electronic schematic drawing was produced on a standard drafting terminal. Electrical components were added from an electronic symbols library.

or chart. By spacing the lines of a fill pattern close together, solid colored fills are possible. This function is extremely useful in producing solid fill patterns for bar graphs, as shown in Figure 11-18.

Another advantage of using CAD for graphs is the variety of lettering styles that can be used to enhance the final plot. Different fill patterns can also be used to set off different parts of a graph. Some plotters make transparencies by using a special pen and loading the plotter with transparent paper. Spline curves can be used for line graphs whose points are connected with a smooth curve. The following is a list of important factors to consider before starting on a graph or chart.

- Sketch the graph or chart.
- Select the paper size and scale.
- Choose the lettering styles for notes and labels.
- Choose fill patterns.
- Assign pen numbers for color graphs and charts.
- Will different layers be necessary for this drawing?
- Think carefully through the steps you will use to draw the graph or chart.

TECHNICAL ILLUSTRATIONS

CAD provides the user a tremendous amount of power to create spectacular graphics. This ability can be used for technical illustrations for advertising, packaging, technical manuals, and so forth. Using 3-D wireframe, surface, or solid modeling software, CAD can create colorful technical illustrations. See the color insert for a few examples.

Orthographic and pictorial illustrations can be created for technical illustrations using virtually the same commands for creating an engineering drawing model. Sometimes the design model can be used for technical illustrations without any major modifications. Sections, fill patterns, exploded assemblies, and different line type and lettering fonts can all be used to create a technical illustration. The power and versatility of most CAD systems makes it an ideal tool for technical illustrations. See Figure 11-19 and the color insert.

OTHER APPLICATIONS OF CAD

There are many other applications of CAD in industry. In some cases special software specifically written for a certain application can be used. Some software is written specifically for civil engineering and mapping applications. Examples of this are shown in the color insert.

One application of CAD that is growing rapidly is for desktop publishing. Many desktop publishing software programs can import graphics created on a CAD system. These CAD files are commonly called DXF files; they can be merged with text created on a word processor through desktop publishing software. The color insert has an example of a technical document being created with a CAD system. After the text is merged with the graphics, the data is output to a laser printer for hard copy. See Figure 11-20.

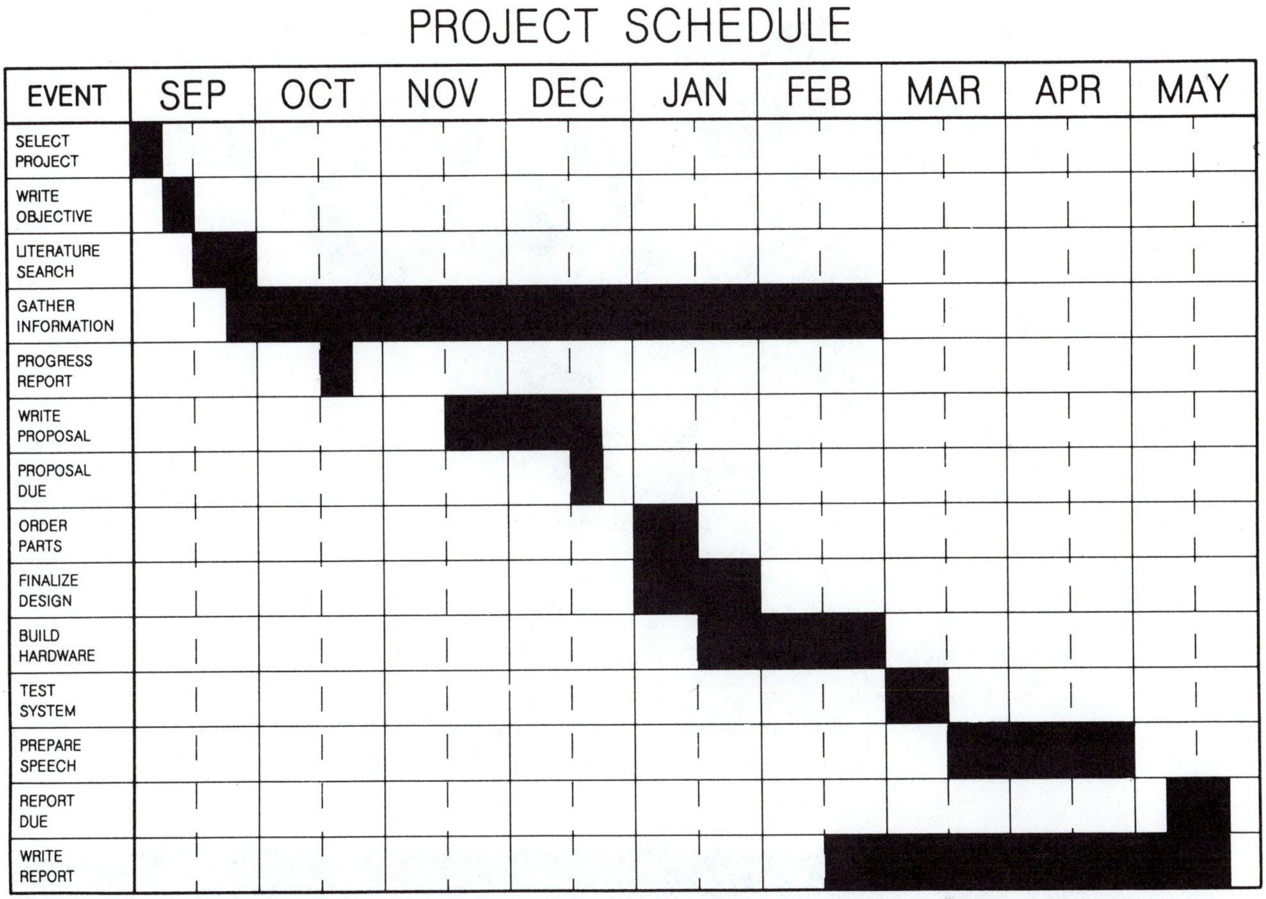

PROJECT SCHEDULE

EVENT	SEP	OCT	NOV	DEC	JAN	FEB	MAR	APR	MAY
SELECT PROJECT	█								
WRITE OBJECTIVE	█								
LITERATURE SEARCH	█	█							
GATHER INFORMATION		█	█	█	█	█			
PROGRESS REPORT		█							
WRITE PROPOSAL			█	█					
PROPOSAL DUE				█					
ORDER PARTS					█				
FINALIZE DESIGN					█	█			
BUILD HARDWARE						█	█		
TEST SYSTEM							█		
PREPARE SPEECH								█	
REPORT DUE								█	█
WRITE REPORT						█	█	█	█

FIGURE 11-18 Bar graph with solid fill patterns created using CAD. *(Courtesy of Hewlett-Packard)*

SUMMARY

The applications of CAD that have been explained thus far in the chapter are the ones that you will most likely be able to cover while in training. However, numerous other applications programs are available to the user, as listed in Chapter Three. Most application programs are very detailed and take additional training time to master. This text is intended only to be used for the initial training of an operator in the fundamentals of CAD. Mechanical, architectural, and electronic drawings, as well as graphs and charts are the most common applications of CAD and can be performed using a standard drafting software package.

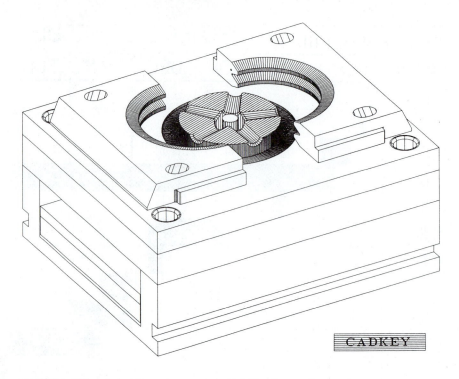

FIGURE 11-19 A technical illustration created with CAD software. *(Courtesy of CADKEY)*

FIGURE 11-20 A desktop publishing workstation showing the computer, laser-jet printer, and image scanner. *(Courtesy of Hewlett-Packard)*

Chapter Eleven GLOSSARY

CAD/CAM—computer-aided design/computer-aided manufacturing is an umbrella process that combines the design and manufacture of a product into one integrated approach through the use of computers.

CAM—computer-aided manufacturing automates manufacturing operations using computer-controlled machines and robots for material handling, and for fabricating, assembling, measuring, and inspecting the product.

CIM—computer-integrated manufacturing is the total automation and computerization of the manufacturing process from receipt of the order to shipment of the completed order.

CNC—computer numerical control is the use of a computer to control some or all operations performed by the numerical control (NC) machine.

DNC—direct numerical control is the use of a computer to provide data to numerical control (NC) or computer numerical control (CNC) machines.

FMS—flexible manufacturing system or cell is a series of computer-controlled machines that perform a complete task and are serviced by computer-controlled transfer devices.

NC—numerical control is a method used to program or provide information to a device to complete a task.

Postprocessor—a computer program used by the processor to change or convert data to a form usable to another device, such as converting graphic data into NC machine language.

Robot—a programmable device used to move or fabricate parts and material.

Chapter Eleven REVIEW

1. List the important steps to be considered before starting on a mechanical drawing.
2. Define CAD/CAM.
3. Define CAM.
4. Define CIM.
5. How can layers be used to simplify architectural drawings?
6. How can colors be used to enhance electronic drawings?
7. How can the COPY command be used in architectural drawings?
8. List the 3 programming methods used with robots.
9. List 2 programming methods used with NC machine tools.
10. Define FMS.

1. Create a solid model of this part.
2. Determine the center of gravity and volume of the model.
3. Create an NC program for fabrication of a part from the solid model.
4. Create orthographic drawings of the model.
5. Create a finite element model of a part and run a finite element analysis.

To accomplish this project, you will refer to Figure 11-21 and create a solid model of the Extension Bar shown in the figure. Use an alloy steel for the material and determine the center of gravity and volume of the part. Create a finite element model of the part. Impose a load on the part and run a finite element analysis of the model. Create orthographic drawings of the part. Create an NC program to turn the Extension Bar from standard round stock.

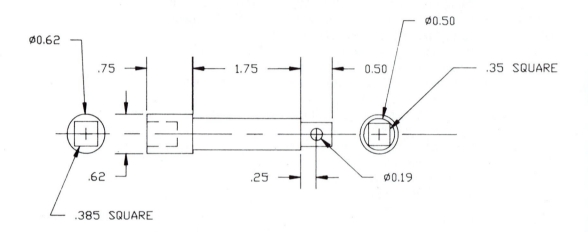

FIGURE 11-21 A short socket extension bar.

Glossary

Alphanumeric—consisting of numbers, letters, and special characters which are input to the computer by means of the keyboard.

ALU—arithmetic logic unit is circuitry that performs the logic and mathematical operations associated with a computer, such as addition and control of data.

Analog computer—uses variations in physical quantities, such as electrical voltage.

Application program—a computer program written for a specific topic, such as architectural drawing or word processing.

Application software—the programmed instructions that allow the user to employ a computer for a specific task or tasks in a format understandable to the operator.

Archive—a process that stores data on magnetic tape which is then stored in a safe place to prevent accidental loss.

Binary number system—a base two number system using ones and zeros to represent information processed by a digital computer.

Bit—a single digit of the binary system having a value of one or zero.

Byte—a group of eight bits.

CAD—computer-aided drafting or computer-aided design.

CAD operator—a drafting technician with CAD training or experience.

CAD/CAM—computer-aided design/computer-aided manufacturing is an umbrella process that combines the design and manufacture of a product into one integrated approach through the use of computers.

CAM—computer-aided manufacturing automates manufacturing operations using computer-controlled machines and robots for material handling, and for fabricating, assembling, measuring, and inspecting the product.

Chip—an integrated circuit etched onto a silicon wafer.

CIM—computer-integrated manufacturing is the total automation and computerization of the manufacturing process from receipt of the order to shipment of the completed order.

CNC—computer numerical control is the use of a computer to control some or all operations performed by the numerical control (NC) machine.

Command—a specific word or phrase used to provide the means for a CAD system to perform a task.

Computer—a device or tool used to process data consisting of input, output, memory, and a central processing unit (CPU).

Computer-aided design—the use of a computer, software, and associated hardware to produce drawings as well as to perform complex engineering functions.

Computer-aided drafting—the use of a computer, software, and associated hardware to produce drawings.

Controller—circuitry used to regulate all operations that take place in the computer. One of the components in a central processing unit (CPU).

Coordinate—a real number used to represent a point in space.

Coordinate geometry—for every point in space a pair of real numbers can be assigned, and for each pair of real numbers there is a unique point in space.

CPU—central processing unit is an integrated circuit (IC) containing the ALU, controller, and part of memory. It is considered to be the "brains" of the computer, and is sometimes referred to as a *microprocessor*.

Crash—mechanical failure in a disk drive or other piece of computer equipment or software program, usually resulting in some loss of stored data.

CRT—cathode ray tube is an output device, similar to a television screen, that is used to display information processed by the computer or entered by the operator.

Cursor—a small flashing line, box or cross hairs, used to locate positions on a CRT.

Data—raw facts and figures represented by such symbols as letters, numbers, or special symbols.

Data base—the handling and storing of information of an entire organization so that it can be universally used by that organization.

Default—a word used to describe the value assigned to certain functions when the computer is turned on. For example, the solid line is the default line used on a CAD system.

Digital computer—a device that uses numbers for the manipulation of data.

Digitizing—the process of identifying, locating or selecting a menu item, entity or point through an input device.

Disk—a circular flat piece of plastic material or aluminum coated with a thin layer of magnetic material used to store data. Sometimes called "floppy disk," "minidisk" or "diskette," "microdisk," "hard disk," "Winchester disk," "disk pack."

Display device—a device used to display a graphic image. The CRT is the most common display device used for graphics.

DNC—direct numerical control is the use of a computer to provide data to numerical control (NC) or computer numerical control (CNC) machines.

Downtime—the period of time associated with a computer breakdown caused by hardware or software problems.

Drum plotter—a pen plotter, electromechanical output device that uses pens mounted in a block that moves across the paper and along a bar with the paper mounted in a rotating drum.

Dual display—the use of two display devices to show graphics and to allow the operator to maintain visual continuity among views, details, layers, and so forth.

Electrostatic plotter—an output plotting device that produces an image by electronically charging the paper and then putting a toner over the charge to produce an image.

Entity—drawing features, such as lines, circles, arcs, and splines.

Existing point—a point located on drawing entities, such as the end points of lines, the center of circles, and the ends of an arc.

Extrusion—a method of creating a 3-D wireframe model by copying a surface and placing it at a new depth.

File—the memory location for data created or input into the computer.

Flatbed plotter—an electromechanical plotting output device where the paper remains stationary on a flatbed while the pens move in two directions creating the plot.

FMS—flexible manufacturing system or cell is a series of computer-controlled machines that perform a complete task and are serviced by computer-controlled transfer devices.

Function—a group of commands that enables a CAD system to perform a specific task, such as drawing circles.

Graphics tablet—an input device used with a puck or a stylus to locate a screen cursor, to digitize a drawing or to select menu items.

Grid—a series of small dots arranged in rows on the screen that can be used for point selection.

Grid point—used as a reference point for drawing entities.

Hardware—all the physical equipment or devices associated with the operation of a computer.

IC—integrated circuit is a microminiaturized circuit etched onto a silicon wafer sometimes referred to as a *chip*. The CPU and circuits used for random access memory (RAM) and read-only memory (ROM) are examples of ICs.

Icon—a picture or graphic representation of a word.

Ink jet plotter—an output plotting device that forces minute droplets of ink onto the paper to produce an image.

Input device—a mechanism used to interact with a computer or to input data.

Interactive—refers to the need for human intervention in the operation of a computer in order to complete a task.

Jaggies—phenomena, associated with raster display terminals, which cause angled lines to appear jagged or stair-stepped. Also known as *aliasing* or *stair-stepping*.

Joystick—an input or locator device used to control the position of a screen cursor.

Kilobyte (K)—a unit of measure for memory storage equivalent to approximately 1000 bytes or 1024 to be exact.

Language—a written set of words or symbols used to communicate with a computer. Languages are identified as BASIC, FORTRAN, PASCAL, COBOL, and others.

Laser plotter—an output plotting device that uses a laser beam and toner to produce an image on paper.

Last referenced point—the position of the pen on the screen, usually represented on screen with a small dot or an X.

Light pen—an input or locator device used to make point and menu selections by sensing light emitted from the picture elements on a CRT.

Locator device—apparatus used to locate or position a screen cursor.

Magnetic tape—a thin strip of plastic film coated with a thin layer of magnetic material used to store data and to archive drawings.

Mainframe computer—a large computer housed in a remote, environmentally controlled setting. Typically, the most powerful type of computer.

Megabyte (Mb)—a unit of memory storage equivalent to approximately 1 million bytes.

Memory—circuitry (random access memory or read-only memory), or media (floppy disk, magnetic tape, and so forth) that stores processed data inside of the computer.

Memory device—a peripheral device used to store programs and data outside of the computer. Examples include magnetic tape, floppy disk, hard disk, and so forth.

Menu—a table or list of drawing and support commands from which a CAD operator can choose to produce a drawing.

Microcomputer—a computer based upon the microprocessor. Strict environmental controls are not needed.

Microprocessor—the miniaturized electronic circuits necessary to process a program; the "brains" of a computer.

Minicomputer—a computer whose processing speeds range between those of the microcomputer and the mainframe computer. Strict environmental controls are needed.

Modem—a device used to send data on telephone lines by MOdulating digital information to audio sound and then DEModulating the audio signal to digital form.

Mouse—an input or locator device used to position a screen cursor.

Multiuser—a CAD system with multiple workstations tied into one CPU with the data base of the design projects available to all users.

NC—numerical control is a method used to program or provide information to a device to complete a task.

Operational software—software developed to control the operation of the computer (CPU) and the peripheral devices. Sometimes called *system software*.

Operator—a person who is trained to use an application software package on a computer.

Origin—the point of intersection of the X and Y axes on Cartesian coordinates. Usually located at the center or bottom left of the CRT on a CAD system. An origin can usually be moved to some other position on the CRT.

Output device—a mechanism used to produce hard copies of images or data created with a computer.

Palette—a range of colors.

Parametric program (macros)—English or programming languagelike statements that call on drawing functions from the application software which are then chained together to perform a task.

Pen plotter—an electromechanical output device that uses pens mounted in blocks or holders that move across the drawing medium to produce a two-dimensional graphic image.

Peripheral—equipment or device, outside of the computer, which is controlled by the CPU.

Photo plotter—an output device that uses a light beam to draw a pattern onto a light-sensitive medium that is developed to produce an extremely accurate drawing or overlay used to make printed circuits.

Pixel—one dot or picture element that makes up part of the raster display terminal.

Polar coordinates—a concept maintaining that any point in space can be identified by stating the angle and the distance to be traversed along the angle.

Postprocessor—a computer program used by the processor to change or convert data to a form usable to another device, such as converting graphic data into NC machine language.

Program—the set of instructions, arranged in a logical sequence, used to command a computer to perform a specific task.

Programmed function board—an input device with rows of buttons that when depressed will activate a specific drawing function, such as a line.

Programmer—a person trained to write programs to control the operation of a computer.

Prompt—a message shown on a CRT to assist the operator of a CAD system in performing a task.

Puck—an input or locator device used with a graphics tablet to position a screen cursor for making menu choices and for digitizing drawings.

Random access memory (RAM)—volatile (transitory) memory used to store data temporarily during processing.

Raster display—a display device that uses thousands of small picture elements that are refreshed thirty to sixty times a second to form an image.

Read-only memory (ROM)—computer instructions that were permanently programmed on a chip during its manufacture.

Refresh—redrawing of a graphic image on a CRT.

Relative Coordinate—reference is made from the last point input.

Resolution—the clarity or sharpness of a display terminal.

Robot—a programmable device used to move or fabricate parts and material.

Software—the chained statements, directions or procedures used by the computer to perform a task.

Solid modeling—a process for designing using combinations of solid geometric primitives, such as blocks, cylinders, cones and spheres.

Standalone—a CAD system with each workstation independent of another.

Stylus—an input or locator device used with a graphics tablet to control cursor location and to select menu items.

Symbol—a graphic representation of a part that is repeatedly used on a CAD system.

Surface Modeling—a graphical modeling technique used to define and describe surfaces.

Third-party software—software developed by a company independent of the hardware manufacturer.

Thumbwheels—an input or locator device used to control the position of a screen cursor, consisting of one wheel which moves the cursor in the X axis and one which moves the cursor in the Y axis.

Turnkey—a word used to describe a CAD system that is installed as a complete package by the manufacturer, and is ready for operation by the purchaser.

Unreferenced point—a point positioned on the screen without regard to other points on the screen.

User friendly—a term used to describe hardware or software that is easily learned and allows an operator to become proficient in a short period of time.

User software—programmed instructions created by the user to enhance or upgrade the applications software. Examples in CAD include user-created symbols and parametric programs.

Vector refresh display—a display device that combines the features of the raster display and the vector storage display, and one which continually refreshes its image and does not suffer from the jaggies.

Vector storage display—a display device that maintains its image for an extended period of time and does not suffer from the jaggies.

Winchester disk drive—a term associated with a hard disk, made of a rigid aluminum platter with a magnetic coating, which spins in an airtight enclosure used to store data.

Window—the current viewing area on screen.

Wireframe—a 3-D model of a part represented by lines giving the object the appearance of being a model made of wires.

X axis—a horizontal line that passes through the origin of the Cartesian coordinates. Values to the left of the origin are negative; values to the right are positive.

Y axis—a vertical line that passes through the origin of the Cartesian coordinates. Values upward from the origin are positive; values downward are negative.

Z axis—a line perpendicular to the X and Y axes intersecting at the origin. Values behind the origin are negative; values in front are positive.

Robo Systems CAD
Chessell-Robocon Corporation
111 Pheasent Run
Newtown, PA 18940
215-968-4422

Appendix A

With the Robo CAD package and a standard Apple II + or IIe computer, you can generate schematics, mechanical drawings, architectural layouts, and business presentations with speed and accuracy previously available only on expensive CAD systems.

The CAD-1 package includes a precision controller, software, user manual, and an interface module which plugs into the games connector on the Apple II computer. The entire system software resides on a single floppy disk.

Your images can be permanently recorded on paper, vellum, drafting film, or photographic slides, for archives or the slide projector. For proof copy, use your dot matrix printer or, for higher quality artwork, choose one of the Robo Systems precision drafting plotters (or other compatible units).

When your design is complete, save it on disk. Modify it if needed, then store it again. Like tape-recorded music, the image remains on the disk until it is purposely erased.

With CAD-2 you "simply point to select" from the on-screen menu and drawing palette. There is instant feedback as your drawing progresses. All changes in line length, arc radius, sweep angle, and more are continuously reported, allowing you to verify point-to-point dimensions as you go.

With CAD-2, all drawing elements are "rubberbanded," changing in size and position as the joystick moves. This is true real-time image manipulation, a great aid in visualizing design changes and graphic effects. You can make as many trial constructions as you wish, then when you have the desired effect on the screen, confirm it with a single push of a button.

RoboCAD-PC

Robo Systems also has software that runs on the IBM PC family of microcomputers. Pull-down menus and moveable icons present the entire set of drafting tools on a single screen. Included with RoboCAD-PC is a translator which converts drawing files to the DXF format to be used with other CAD systems. DXF files from other systems can also be translated for use with RoboCAD-PC.

SECTION 5

APPENDIX

A CROSS SECTION OF CAD VENDORS

Appendix A is included to offer information about the different kinds of CAD systems on the market. The system on which you are being trained probably will not be the same system that you will be working with. This section has been included to acquaint you with the different types of systems and to describe them for a better understanding of CAD systems. In addition, it is important to stay current with the latest technology in CAD. This is a very dynamic field, so additional information can be obtained from the manufacturers whose addresses and phone numbers are provided with each product description. This list of CAD vendors is not meant to be all inclusive, but merely a typical cross section.

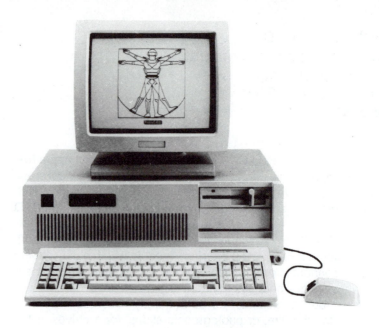

FIGURE A-1 RoboCAD PC. *(Courtesy of Robo Systems.)*

AutoCAD
Autodesk, Inc.
2320 Marinship Way
Sausalito, CA 94965
415-332-2344

AutoCAD allows anyone who draws to create and revise drawings onscreen, and to store their work for later use. You can move, copy, mirror, rotate, erase, stretch, trim, extend, scale, and dimension any part of a drawing or the entire drawing-among other options. Arrays of objects, such as bricks or gear teeth, can be generated automatically.

You can annotate drawings with text of any size, position, and angle, choosing from a variety of text fonts and styles. And when you are ready, your drawings can be plotted to any size and scale.

You create and revise drawings onscreen using a digitizer or mouse to point to screen or tablet menu items, and to move the drawing cursor around the screen. Once you've stored your drawings on a floppy or hard disk, you can plot the drawings at any scale, or use components of drawings in other drawings.

You can position drawing elements on the screen by freehand pointing with the cursor, or by typing in coordinates. Virtually all AutoCAD commands dynamically display their results. When you move an object in a drawing, for example, you can drag the object to the desired location and place it with the push of a button.

Anything you do with AutoCAD you can also undo, whether you are erasing, moving, copying, breaking, cross-hatching, stretching, or rotating objects in a drawing. You can even undo all the way back to the beginning of an editing session if you need to.

Many AutoCAD users have developed add-on programs that work with AutoCAD, for applications ranging from architecture to electrical and medical engineering, fire and security systems, and even orthodontics. More than 250 third-party applications are available through the AutoCAD Applications Catalog.

AutoCAD is programmable using AutoLISP. AutoLISP facilitates the development of applications and utilities that further ease your work. AutoCAD runs on most microcomputers supporting the PC-DOS and MS-DOS operating systems, on the IBM RT PC under AIX, and on the Sun Microsystems family of Sun-2 and Sun-3 32-bit technical workstations under UNIX. AutoCAD supports more than 140 peripheral devices for input and output.

AutoCAD drawings can be exchanged with other applications software, databases, and mainframe CAD systems. AutoCAD supports the Initial Graphics Exchange Standard (IGES), so drawings from most CAD systems can be accurately and easily translated to and from AutoCAD.

Design professionals worldwide use AutoCAD—more than they use any other CAD software—for uses as varied as architecture, civil, electrical, and mechanical engineering, schematics, flowcharts, system design, shopfitting, graphic design, factory and facilities planning, technical illustration, office layout, printed circuit board design, numerically controlled machine programming, archaeological site documentation, and theatrical lighting.

AutoCAD AEC

AutoCAD AEC is a powerful design and drafting software package for professionals in architecture, engineering, and construction. The software comes with symbols to simplify some work. AutoCAD AEC automatically generates reflected ceiling plans, as well as mechanical, electrical, and other background drawings. Once drawn, these can be copied, resized, mirrored, or revised without losing the original drawings.

The program contains a complete library of shapes and symbols. Structural, plumbing, electrical, furniture and appliances, site planning, and titling symbols are all available for automatic insertion anywhere in drawings. As the drawing is created, a working database is created. The specifications for all the door, window, and plumbing fixtures schedules are embedded in the drawing themselves and can be easily extracted.

DiscoverCAD
Hearlihy & Company
714 W. Columbia
P.O. Box 869
Springfield, OH 45501
800-622-1000

DiscoverCAD uses easy-to-read pull-down menus for command selection. Command and coordinate entry is accomplished using the AppleMouse. DiscoverCAD features automatic dimensioning: leader lines, arrowheads, and dimension text are automatically drawn, using linear or radial dimensioning options. You can easily and quickly zoom in and out or pan the screen for added flexibility.

DiscoverCAD features 128 layers which can independently be turned on or off. Prints can be produced on virtually any parallel or serial dot matrix printer. Plots can be produced on Houston Instruments or Hewlett Packard plotters. This software requires an Apple IIe or IIc with 128K of memory, two 5¼" disk drives, and an AppleMouse.

CADDRAW
Hearlihy and Company

This software is a low-cost program used to teach entry level CAD. CADDRAW features allow you to automatically center any simple multiview or isometric drawing on any one of four title blocks and borders. It also has a simplicity lock that can be used to turn off complex operations to keep it simple for the first-time user. CADDRAW can scroll from page 1 or 2, layer both pages, or save shapes from one page to be used as symbols on another.

CADDRAW is available for the Apple II+, IIe, or IIc with 48K of memory required. CADDRAW can be used with a number of different input devices such as mice, light pens, joysticks, and tablets. Symbols can be created and added to the symbol tables provided with the software. Symbols can also be combined, or edited. CADDRAW can be used for 2-D or 3-D landscaping, architecture, mechanical, or electronic drawing symbols.

Hard copy prints can be produced on virtually any dot matrix graphic printer. Prints can be made to scale and cropped, rotated, reversed, magnified, and printed in a variety of densities.

FIGURE A-2

VersaCAD Designer
7372 Prince Drive
Huntington Beach, CA 92647
714-847-9960

VersaCAD Designer provides 2-D drafting, 3-D modeling, detailed report generation, and CAD communications together in a single software package. The program supports many different computers and peripherals. The software can be modified by the user, providing tools for a personal design system, including making your own menus, recording and playing back design operations through macros, defining screen colors, a built-in CAD programming language called CPL.

Color shading in 3-D is made possible by a user-defined light source, high resolution (1024 by 1024 pixels), and a range of 256 colors. There are built-in primitives in both 2-D and 3-D such as spheres, cylinders, polyhedrons, curves, and cones. Two-way transfer between 2-D and 3-D can also be done. Models can be displayed in orthographic, wireframe, isometric, or perspective views with hidden line removal.

Presentation graphics capabilities include film recorder and laser printer output, and off-line plotting. Universal communications in both directions with other CAD systems is built in, including an IGEX standard translator. Bill of materials report capabilities are included which provides automatic sorting of drawing parts, calculating, and report generation.

You can add mechanical engineering and architectural modules, CAM interface or an interactive database for a complete desktop design station.

FIGURE A-3   Microcomputer-based CAD workstation. *(Courtesy of VersaCAD)*

CADKEY
Micro Control System, Inc.
27 Hartford Turnpike
Vernon, CT 06066
203-647-0220

CADKEY is a true 3-dimensional design and drafting system for the IBM PC family of microcomputers. Applications range from mechanical engineering, design, drafting, analysis, and manufacturing to technical illustrations, graphic arts and biomedical engineering.

With a 3-D data base, CADKEY is able to integrate 2-D drafting and 3-D design with external programs for numerical control, finite element analysis, bill of materials, and more. CADKEY can also be customized to meet individual needs.

With Version 3.0, designers have complete design integration on their personal computer with one fully integrated PC-CADD system. The CADKEY Advanced Design Language (CADL) includes 3-D data primitives, high level CADKEY command access, unlimited program size, and macro tablet management.

Surface meshing includes polygons (with graphics fill up to 256 colors) polylines (for continuous numerical control tool paths), points, and lines. CADKEY allows the user to automatically generate an unlimited number of views from the 3-D model. Changing geometric entities in one view will update all the others.

The Immediate Mode Command set is a group of more than two dozen commands that can be accessed at any time within the program without having to leave the current command. An IGES translator can be used to bidirectionally move files between different CAD systems.

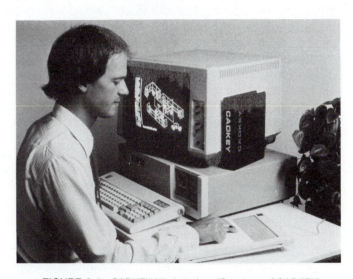

FIGURE A-4   CADKEY Workstation. *(Courtesy of CADKEY)*

Appendix B

PURCHASING A CAD SYSTEM—VENDOR QUESTIONNAIRE

Evaluating a CAD system for purchase can be a very time-consuming task because of the more than 100 CAD systems on the market. Looking at every system would involve countless hours, becoming an impossible task. However, not taking the necessary time to study one's needs and carefully evaluate the systems most suited to those needs would be a mistake. Purchasing a system hurriedly will likely lead to regret and dissatisfaction with the system purchased. Some broad guidelines can be followed to ensure that the CAD system purchased will meet current and future needs, and perform the functions that are important to all concerned.

A good place to begin your evaluation is to educate yourself on CAD. Become familiar with the terminology associated with CAD. Do not start talking with vendors right away. Numerous regional and national conferences on CAD are conducted periodically throughout the year. Attending one of these conferences is an excellent method of becoming more familiar with CAD. Some of these conferences are listed in Appendix C.

As you become more knowledgeable about CAD, it may then be the time to start collecting information on the numerous systems on the market. Trade journals, tool shows, computer shows, and national conferences all provide information related to CAD systems. Books listing CAD vendors are also excellent sources for obtaining CAD information. Trade journals and books of vendors are listed in Appendix C.

After you have become familiar with CAD, it is time to analyze your needs. Consider the needs of the entire company. Look into the future through long-range planning to ensure that the system purchased fits in with the outlook. Estimate the amount of expansion that may be required of the system. List the applications programs that a CAD system should have to meet your needs.

Because of the preliminary needs assessments, you should be able to narrow down the field of CAD systems significantly, possibly to just a single group. For example, you may find that to fulfill your needs it is necessary to evaluate only mainframe-based CAD systems. This still leaves a number of those systems to be evaluated, but many of them can be eliminated on the basis of cost, application software available, and hardware needs.

Once the number of systems has been reduced, you can begin evaluating vendor demonstrations. To help this evaluation process and to ask the right questions, you may want to bring a questionnaire to the demonstration. A sample questionnaire is shown at the end of this Appendix. It is also important that drawings common to your company be brought and drawn on the CAD system. After the demonstration, sit down and try the system. A trained drafter should be able to determine if a system is easy to use and "friendly."

Other concerns are the credibility of the vendor, training, maintenance, and software updates. Because of the dynamic nature of CAD and computers, some companies will not be able to compete or make market adjustments, which could cause bankruptcy. Make sure that there is adequate training available for the operators. Guaranteed 24-hour service is a must to prevent downtime and loss of money for the company. Find out the cost of the maintenance contract and exactly what it covers. Finally, determine the number of software updates and enhancements that can be expected.

This may seem to be an expensive and long-term method of purchasing a CAD system. But following these guidelines will help the purchaser to meet needs and to get the most out of a CAD system. Choosing the right system will be a cost-effective investment that will pay for itself many times over in increased productivity, better designs, and a more competitive product.

CAD Vendor Questionnaire

Manufacturer_____

Model_____

Turnkey System?   Y/N

Hardware

1. Can you link multiple systems?   Y/N

2. Maximum number of workstations per CPU_____

3. Standalone system?   Y/N

4. Standard CRT size_____

    Storage, Raster, or Refresh?_____

    Screen Color (i.e., green)_____

    Resolution_____

Optional screen size_____

Are color CRTs available?   Y/N

5. Are prompts displayed on the working CRT or on another screen?

_____

6. Any environmental restrictions?_____

Peripherals:

1. Methods of input? (i.e., light pen, puck, mouse, etc.)_____
2. Drawing storage medium_____

   Maximum number of drawings stored on medium_____
3. Hard copy units available?   Y/N
4. Plotters: Make, model, paper size & number of pens: _____

   _____
5. Digitizer & sizes_____
6. Can the system handle more than one operation at a time?   Y/N

   If so, what?_____

   Is a tape reader/punch available?_____

Maintenance

1. Warranty period_____
2. Single source for hardware & software?   Y/N

   Nearest service to_____
3. What is guaranteed service response time?_____
4. Cost of maintenance contract?_____

   Does this include software enhancements?_____
5. Is there a toll-free hot line for assistance or questions?_____

Software

1. Language of graphics source?_____

2. Is user programming available to customize software?_____

3. Does the menu appear on screen or graphics tablet?_____

4. How are updates and enhancements handled?_____

_____

5. Is there a symbols library?   Y/N

   What does it include?_____

   _____

6. 3-D Drafting capabilities?   Y/N   Cost?_____ Wireframe_____

7. English and metric?   Y/N   Surface _____

8. CAD/CAM capabilities?   Y/N   Cost?_____ Solid _____

9. Engineering analysis programs (e.g., finite element)?   Y/N

10. Any type of automatic generation of bill of materials?   Y/N

    Cost?_____

11. If not available now, which of these programs are planned in the next

    two years?_____

    _____

12. What major enhancements have been made within the last year?

    _____

    _____

13. Self-instruction program or help command to assist new opera-
    tor?   Y/N

14. Parametric programming language?   Y/N

15. Automatic dimensioning?   Y/N

16. Line types available: _____

    _____

17. Lettering styles: _____

    _____

18. Crosshatch patterns—*user defined, symbols, both*

19. Can section of drawings be moved?   Y/N

20. Can you rotate?   Y/N   scale?   Y/N   mirror?   Y/N

21. Number of levels?_____

22. IGES translator _____

Training

1. Where is training available?_____

   Cost?_____

2. Length of training?_____

   Number of days?_____   Number of persons?_____

3. Would a demonstration be possible at our site?   Y/N

4. Follow-up visit after installation?   Y/N

5. Training guide available to user?   Y/N

Users' Group

1. Is there a formal users' group?   Y/N

2. How often does it meet?_____

3. Where does it meet?_____

4. Is there a list of users available?_____

5. Is there a library of user programs?_____

Miscellaneous

1. Delivery time?_____

2. What enhancements are you planning in the next year?_____

   _____

   _____

Appendix C

CAD/CAM DIRECTORIES, PERIODICALS, CONFERENCES

The following directories, periodicals, and conferences are sources of additional CAD/CAM information.

Directories

Computer graphics: U.S. Directory of Vendors
Daratech, Inc.
16 Myrtle Ave.
P.O. Box 410
Cambridge, MA 02138

The S. Klein Directory of Computer Graphics Suppliers
Technology & Business Communications, Inc.
730 Boston Post Road
P.O. Box 392
Sudbury, MA 01766

Low cost CAD systems and a survey of CAD/CAM systems.
Leading Edge Publishing
11551 Forest Central Drive
Dallas, TX 75205

Periodicals

The Anderson Report
Automation News
CIM Technology
Commline
Computer Aided Engineering
Computer Graphics News
Computer Graphics World
Design Graphics World
S. Klein Newsletter on Computer Graphics
Machine Design
CAD/CAM Alert
S. Klein Computer Graphics Review

Organizations Sponsoring CAD or CAD/CAM Conferences

Society of Manufacturing Engineers
    P.O. Box 930
    Dearborn, MI 48128   Phone 313-271-1500
NCGA Seminars (National Computer Graphics Association)
    2033 M Street N.W., Suite 300
    Washington, DC 20036   Phone 202-466-4102
Center for Manufacturing Technology
    4170 Crossgate Drive
    Cincinnati, OH 45236   Phone 513-791-8801
Frost & Sullivan, Inc.
    106 Fulton Street
    New York, NY 10038   Phone 212-233-1080
American Institute for Design & Drafting
    3119 Price Road
    Bartlesville, OK 74003   Phone 918-333-1053
American Society of Mechanical Engineers
    345 East 47th Street
    New York, NY 10017   Phone 212-644-7100

Appendix D

REFERENCES AND SUGGESTED ADDITIONAL READINGS

Ayres, R. V. *The Impact of Robotics on the Workforce and Workplace.* Pittsburgh: Carnegie-Mellon University, June 1981.

Bylinski, Gene. A New Industrial Revolution Is on the Way. *Fortune,* October 5, 1981, 106–114.

CAD/CAM: Management Strategies. Auerbach Publishers, 1984, 350 pages.

The CAD/CAM Handbook. Bedford, MA: Computervision Corporation, 1980.

Coiffet, P. *An Introduction to Robot Technology.* New York: McGraw-Hill, 1983.

Computer-Aided Design, Engineering, and Drafting. Auerbach Publishers, 1984, 300 pages.

Demel, J. T. & M. J. Miller. *Introduction to Computer Graphics.* Monterey, CA: Brooks/Cole Engineering Division, 1984.

Dorf, R. C. *Robotics and Automated Manufacturing.* Reston, VA: Reston, 1983.

Feigenbaum, E. A. *The Fifth Generation.* Reading, MA: Addison-Wesley, 1983.

Finkel, J. I. High Tech Displays Get Lower Price Tags. *Computer-Aided Engineering,* July/August 1983, 32–41.

Foley, J. D. & A. Van Dam. *Fundamentals of Computer Graphics.* Addison-Wesley, 1982.

The Future of Solids Modeling. *Machine Design,* September 22, 1983, 107–108.

Getting Started on Acquiring CAD. *Commline,* March–April 1984, 30–34.

Gilio, W. K. *Interactive Computer Graphics.* Englewood Cliffs, NJ: Prentice-Hall, 1978.

Golden, F. Big Dimwits and Little Geniuses. *Time magazine,* January 3, 1983, 30–32.

Groover, M. P. & E. W. Zimmers, Jr. *CAD/CAM: Computer-aided Design and Manufacturing.* Englewood Cliffs, NJ: Prentice-Hall, 1984.

Gunn, T. G. *Computer Applications in Manufacturing.* New York: Industrial Press, 1981.

Halfhill, T. R. Mass Memory Now and in the Future. *Compute,* March 1983, 54–65.

Hartman, R. A., et al. *Computer Graphics,* College Station, TX: Creative Publishing, 1983.

Here Come the Robots. *Newsweek,* August 9, 1982, 58.

Hopper, G. M. & S. L. Mandell. *Understanding Computers,* St Paul, MN: West Publishing, 1984.

Hudson, C. A. Computers in Manufacturing. *Science,* February 12, 1982, 818–825.

Ingelsby, T. CAD/CAM: Should We... *Assembly Engineering.* March 1982, 48–50.

Ingelsby, T. The Many Sides of CAD/CAM. *Assembly Engineering.* October 1982, 10–15.

Japan's Bid to Outdesign the US. *Business Week.* April 13, 1981, 123–124.

Jobs. A Million That Will Never Come Back. *US News & World Report.* September 13, 1982, 53–56.

Kilburn, P. & E. Teicholz. Low Cost CAD at Work. *Datamation.* January 1983, 103–110.

Koren, Y. *Computer Control of Manufacturing Systems.* New York: McGraw-Hill, 1983.

Krouse, J. K. *What Every Engineer Should Know About Computer-aided Design and Computer-aided Manufacturing.* New York: Decker, 1982.

Krouse, J. K. Smart Robots for CAD/CAM. *Machine Design,* June 25, 1981, 85–91.

Krouse, J. K. CAD/CAM Broadens Its Appeal. *Machine Design,* April 22, 1982, 54–59.

Krouse, J. K. Reducing the Labor Content in Finite Element Analysis. *Machine Design.* September 8, 1983, 64–69.

Krouse, J. K. Selecting a Graphic Input Device for CAD/CAM. *Machine Design,* October 6, 1983, 75–80.

Mufti, A. A. *Elementary Computer Graphics.* Reston, VA: Reston, 1983.

Muralijacic, T. Computer Graphics: A 20-year Forecast. *Journal of Systems Management.* November, 1981, 17–21.

Newman, W. M. & R. F. Sproull. *Principles of Interactive Computer Graphics.* New York: McGraw-Hill, 1973.

Pressman, R. S. & J. E. Williams. *Numerical Control and Computer-aided Manufacturing.* New York: Wiley, 1983.

Reed, J. S. Computer Graphics. *Mini-Micro Systems,* December, 1982, 210–221.

Retraining Displaced Workers: Too Little, Too Late? *Business Week,* July 19, 1982, 178–185.

Stein, Kathleen. Alice's Factory. *Omni,* April 1982, 44–49.

Teresko, J. CAD/CAM Goes to Work. *Industry Week,* February 7, 1983, 40–47.

Thornburg, D. The Fifth Generation. *Compute,* July 1983, 18–20.

Tver, D. F. *Robotics Sourcebook and Dictionary.* New York: Industrial Press, 1983.

Whittier, R. J. Semiconductor Memory. *Mini-Micro Systems,* December 1982, 188–196.

INDEX